The Smell of Kerosene:
A Test Pilot's Odyssey

The Smell of Kerosene
A Test Pilot's Odyssey

The Smell of Kerosene tells the dramatic story of a NASA research pilot who logged over 11,000 flight hours in more than 125 types of aircraft. Donald Mallick gives the reader fascinating first-hand descriptions of his early naval flight training, carrier operations, and his research flying career with NASA and its predecessor agency, the National Advisory Committee for Aeronautics (NACA).

Mallick joined the NACA as a research pilot at the Langley Memorial Aeronautical Laboratory at Hampton, Virginia, where he flew modified helicopters and jets, and witnessed the NACA's evolution into the National Aeronautics and Space Administration.

After transferring to the NASA Flight Research Center (now NASA Dryden Flight Research Center) at Edwards, California, he became involved with projects that further pushed the boundaries of aerospace technology. These included the giant delta-winged XB-70 supersonic research airplane, the wingless M2-F1 lifting body vehicle, and the triple-sonic YF-12 Blackbird. Mallick also test flew the Lunar Landing Research Vehicle (LLRV) and helped develop techniques used in training astronauts to land on the Moon.

This book puts the reader in the pilot's seat for a "day at the office" unlike any other.

NASA SP 4108

The Smell of Kerosene:
A Test Pilot's Odyssey

By Donald L. Mallick
with Peter W. Merlin

The NASA History Series

National Aeronautics and Space Administration
NASA History Office
Washington, D.C. 20546   2003

Library of Congress Cataloging-in-Publication Data

Mallick, Donald L., 1930-
The smell of kerosene: a test pilot's odyssey/by Donald L. Mallick with Peter W. Merlin.
p. cm. — (The NASA history series) (NASA-SP 2003-4108)
Includes bibliographical references and index.
1. Mallick, Donald L., 1930- 2. Test pilots—United States—Biography. I. Merlin, Peter W., 1964- II. Title. III. Series. IV. NASA SP 4108

TL540.M333A3 2003
629.134'53'092—dc22
[B]   2003060745

Dedication

For my brother, Robert "Bob" Mallick, who taught me how to swim and whose Army Air Corps Service in World War II inspired my career in aeronautics.

Contents

Prologue: Mach 3+ ix

Acknowledgements xi

Chapter 1: Wings of Gold 2

Chapter 2: Naval Air Operations 20

Chapter 3: NACA and NASA – Langley, Virginia 43

Chapter 4: Jet Research at Langley 58

Chapter 5: Super Crusader 79

Chapter 6: High Desert Flight Research - Edwards, California 97

Chapter 7: The Best of Times, The Worst of Times 122

Chapter 8: Flying the Heavies 146

Chapter 9: Triple-sonic Blackbird 168

Chapter 10: A Time of Transition 200

Chapter 11: Shifting Gears 221

Epilogue 232

Glossary 233

Appendix A: U.S. Navy Aircraft Flown (1950-1963) 238

Appendix B: Flight Research Programs At Langley (1957-1963) 239

Appendix C: Aircraft Flown While At DFRC (1963-1987) 240

Appendix D: Published Technical Papers 242

Index 244

Prologue

Mach 3+

Outside the cockpit, the hush of the morning is shattered. It sounds like a cross between a chainsaw and someone battering a manhole-cover with a jackhammer. I'm glad that I'm inside the cockpit, like a titanium cocoon with quartz glass windows. My pressure suit helmet muffles the clatter of the "start cart" and allows me to hear the gentle hiss of oxygen.

I continue to run through the pre-start checklist with my backseater. The crew chief signals that the starting unit is connected. As I ignite the left engine, observers behind the aircraft see a bright green flash illuminate the exhaust duct. The engine roars to life. I repeat the procedure for the right engine. The roar outside must be deafening. I can see that the ground crew is wearing ear protectors, but a few are also holding their hands tightly to the sides of their heads.

I finish my checklist and verify that the flight controls are functioning properly. We have been in the cockpit now for nearly three-quarters of an hour. I release the brakes and ease the throttle forward. The sleek jet rolls onto the taxiway, its black skin soaking up the sunlight. I imagine that it looks like a sinister shadow slipping toward the runway.

I line up for takeoff squarely on runway centerline, with nearly three miles of concrete in front of me. As I push the throttles forward, the afterburners ignite almost simultaneously. I advance the throttles to maximum and the engines pour blue fire out through the exhaust ejectors. The Blackbird eats up runway as I let the speed build. At 210 knots indicated airspeed, I pull back gently on the stick and separate the Blackbird and myself from Mother Earth.

Climbing through 10,000 feet, I accelerate to about 400 knots. The pale blue sky rapidly darkens as I climb through 20,000 feet at nearly the speed of sound. At about 34,000 feet, I push the nose over gently, controlling my angle of attack during transition through a critical, high-drag region of the flight envelope. Having established a constant velocity of 450 knots (about Mach 1.2 at this altitude), I begin a climb to cruise altitude.

Racing toward 70,000 feet, I notice the curvature of the Earth, rimmed in azure. My part of the sky is so dark now that it is almost black. Desert and mountains stretch out below. I can see the city of Las Vegas, the Colorado River, and myriad dry lakebeds dotting the landscape. Soon, having reached cruise altitude I watch my airspeed indicator pass Mach 3.0 (more than 2,000 miles per hour).

The air outside is freezing cold, but its friction heats the surface of my airplane to more than 400°F. Insulation and the air conditioning system keep the cockpit a comfortable 60°F, and I am further protected by my bulky pressure suit.

There is no more time for sightseeing. I turn north, toward the Canadian border. In a few minutes, I will begin to work my test points and gather data. Glancing once more out the window, I feel a sense of wonder. How does a small town Pennsylvania boy like me now find myself flying near the edge of space?

Acknowledgements

My deepest thanks to Michael Gorn, chief historian at NASA Dryden Flight Research Center, for his support and encouragement. Dryden *X-Press* editor Jay Levine outdid himself in designing the page layout of this book. Thanks to the Dryden Photo Office for providing some of the outstanding pictures used to illustrate the story. Special thanks to Brian Nicklas of the National Air and Space Museum archives for supplying hard to find historic images of NACA airplanes at Langley Research Center. Christian Gelzer, Betty Love, Sarah Merlin, and Curtis Peebles were kind enough to proofread the text. Steve Lighthill helped shepherd the book through the publishing process.

odyssey – n., a long journey containing a lot of exciting or dangerous events

1
Wings of Gold

Roots – Western Pennsylvania

My roots lie deep in the steel country of America's heartland. My father was a native of western Pennsylvania and my mother had immigrated from Magedeburg, Germany, in 1904. They were married in 1922. My brother Bob was born the following year. They lived in a number of small towns in western Pennsylvania while my father completed his apprenticeship as a machinist. They eventually settled on Neville Island when my dad went to work for the Dravo Corporation. I was born on 4 October 1930 in nearby Sewickley, Pennsylvania.

Neville Island sits in the middle of the Ohio River, about 10 miles northwest of downtown Pittsburgh. Known as "The Point," this is where the Allegheny and Monongahela rivers join together to form the headwaters of the Ohio River. The eastern half of the island was dominated by the steel industry and the western half of the island, where my home was located, was primarily residential.

Living in the middle of the Ohio River provided wonderful opportunities for recreation and my family took advantage of it, especially in the summer. My brother had the honor of teaching me to swim in the Ohio. He used a very direct method of instruction. When I was nine years old, Bob supported me as he waded out to a spot where a submerged rock enabled me to stand with my head just above water. This was about fifty feet from the shore. Bob then left me and swam to the beach and waited for me to swim to shore. I can remember treading the water with my hands to maintain my balance on the rock against the river current until I had the courage to strike out for shore. After a short time, I decided my brother was not coming back out for me so I left my rock and struggled to shore. I'm not sure if it was a dog paddle, a crawl, or a breaststroke, but there was a lot of splashing water as I made slow progress toward the beach. Finally, I made it. From there on it was fairly easy. I had gained confidence that I could keep my head above water and move from one place to another under my own power.

I look back on my early school years with fondness. My parents encouraged me to excel in my classes. I suspect this was due to the fact that when they were children, their opportunities in school were limited and they both entered the "work force" at a very young age. With their encouragement, I applied myself and did reasonably well in my studies.

When the United States entered World War II, Pittsburgh began to gear up for the war effort. Workers manned the mills and factories seven days a week

Young Don Mallick sitting on a pony in 1937.

Courtesy of Don Mallick

for three shifts a day. The furnaces and machines in the factories operated continuously.

Everyone in our family expected my brother Bob, then a senior in high school, to enter military service. This may have weighed heavily on my dad as he contemplated his own experience in Europe as an infantryman during World War I. One of the reasons my grandmother left Germany in 1902 and immigrated to the United States was to keep her sons out of the German army. She succeeded with her sons, but her grandsons were now going to enter the U.S. military. When Bob graduated in 1942, he enlisted in the Army Air Corps as an aviation cadet.

Like millions of Americans with loved ones in the military our family spent the next four years working and praying for victory and the end of the war. For my part, I gathered "pig iron" from slag piles along the river and sold it to a local junk dealer. I invested my profits in Savings Stamps and Bonds. I felt that this, and working hard in school, were my contributions to help the war effort.

Bob completed flight training and earned his wings and commission in the U.S. Army Air Corps. Following assignment to fly B-24 bombers in the 8th Air Force, he went to England. Our family never worried more than during the months that he flew his 25 combat missions in Europe. Crews and airplanes were shot down daily and we knew the dangers that Bob faced. Many of the blue stars that appeared in the windows of local homes signifying that a family member was in the service were replaced with gold ones signifying that the serviceman had given his life for his country. We prayed everyday that our star would stay blue. Thanks to the good Lord, Bob made it through safely. He finished his tour shortly before the end of the war in Europe. His next assignment involved transition training in B-29s. Fortunately, the war with Japan ended before Bob was sent to the Pacific.

After the war, my parents encouraged my brother to go to college, but he was married and wanted to find work. After several jobs in steel mills, he found a job with a local electric utility company. As a sophomore in high school, I had only short-term goals. I planned to get a job after graduation and make a little money. I figured that my father and brother had done very well with a limited education. The Dravo Corporation promoted my dad to Master Machinist and my brother prospered with the power company. My parents still wanted one son to go to college, but I wasn't sure that I wanted to fill that slot. I did not send in any college applications even though my mother had a number of them around the house.

A month after graduation, I had a job in the steel industry and I was on my way to earning the "big money." For three months, I worked the midnight shift in a steel forging plant that made automobile parts. It was an extremely tough job, often leaving me with burns on my hands and arms. Soon, I realized that my parents had the right idea. It was time to send in those college applications.

Penn State

In high school I enjoyed my math and science courses the most, so I decided to study mechanical engineering in college. My dad tried to help me through the company he worked for, Dravo Corporation. A company representative reviewed my high school records and offered to advance all tuition and costs at a low-interest or no-interest loan with a guaranteed position at Dravo when I graduated. He indicated that they preferred Lehigh University. It was a good offer, but it wasn't what I wanted. I had found a new focus for my future, in aviation. The military required a minimum two-year college education for aviation cadets. I decided that I would enter military flight training when I completed the required two years of college. My dad agreed to help me attend Pennsylvania State College. The tuition at the State College was lower and the school had a fine reputation. I thanked the Director of Dravo for his offer and he wished me well.

I applied for and was accepted at a Penn State center for my freshman year. In 1948, the colleges were still loaded with World War II veterans and Penn State did not take any freshmen on the main campus. They were "farmed out" to state teachers colleges (four-year schools, primarily for teachers) and centers. The center that I attended was located on a large country estate that had been donated to the Pennsylvania State College system by a lumber baron named Behrend. The next year, when I went to the main campus at Penn State, I found the other students referred to the Behrend students as coming from "the country club." With only 150 students on this beautiful estate, it looked like a country club setting, but it was a place of serious study. All of the courses were from the Penn State curriculum and the majority of teachers came from the main campus.

During my sophomore year, I roomed with a high school friend, Howard McCullough, in one of the dormitories. The main campus was very impressive and accommodated over 7,000 students. The regular classes had 40 to 50 students and the lecture classes sometimes had more than 300. I felt fortunate that I was rooming with Howard since we were both studying Engineering. After a month or two in this second year of college, I reaffirmed my decision that I would finish this sophomore year and then enlist in the Air Force flight training program. The Army Air Corps that my brother proudly served in had become the Air Force in 1947. The war in Korea was just beginning in the spring of 1950. Howard had joined the local Army National Guard unit with the hopes that he might be able to finish college and get his degree. He encouraged me to do the same, but I had other plans for flight training.

During the spring semester, I met a student who was a Naval Aviator, who had recently finished his four-year tour. He was out of the Navy and back in college to get his engineering degree. We became friends and he filled me with glorious tales of naval aviation. It sounded great, although I wasn't too sure about the carrier landing operation that he described. That sounded a little tough.

Joining the Navy

When the 1950 spring semester was over, I returned home to Neville Island. I wasted no time, swiftly heading to the recruiting office in Pittsburgh. First, I went to the Air Force recruiter and tried to sign up. He asked a few questions and told me that I was too young for the U.S. Air Force. They required their cadets to be 20.5 years old and I was a year shy of that. I had hoped to serve in the same service that my brother did, but I wasn't willing to wait another year. I wanted to fly. I walked down the hall to the Navy recruiter and introduced myself. He put his arm on my shoulder and said, "Welcome, sit right here, I have a few little tests for you." As I took each test, he graded it, smiled, and handed me another. In about one hour, I had passed all the local exams and had a train ticket and schedule in my hand. I then proceeded to the Navy facility at Akron, Ohio, to take a pre-induction physical.

I passed the physical with the stipulation that I have some dental work completed before reporting to Pensacola, Florida, for basic training. This was not a problem, because my reporting date was in September. I had my summer job lined up and I planned one more summer at home, before going into the service. I did not tell my parents about my enlistment plans until I had been accepted. My brother knew that I was applying and he was pleased to learn that I had made it. Now it was time to break the news to my parents. They really wanted me to stay in college and finish my studies. I waited until the evening, when my brother was there to give me a little moral support. When I told my folks, their faces dropped and they were obviously upset. I turned to my brother for support and he, too, started to give me hell for dropping out of college. A little later when the dust had settled, my brother joined me outside and said, "You did the right thing, I just made that scene with mom and dad to take a little heat off of you."

I anticipated a great summer with a job, a little dating and fun, and then off to naval flight training in September. The summer job at the high school prevented me from becoming bored and it provided a little cash. The war in Korea was heating up and a number of my high school friends were in the service. One of them, Jim Wiley, came home from the Navy on two weeks leave. We "bumped around" together, drank a few beers, and saw the sights. A few days before Jim was to report back, I asked him if he wanted to set up a double date and have dinner. I offered to set him up with one of the girls I had been dating but he indicated that he knew a nice gal from Coraopolis that he would call.

I'm glad that he did. That gal, Audrey Waite, later became my wife. Audrey was attractive with a super personality and I was very impressed when I met her. Jim had to report back to duty and a few days later I called Audrey and we began seeing each other. Audrey was a great dancer and a nice person to be with. It was obvious to me that she stood out head and shoulders above the other girls that I had been dating. I was dismayed that I had not met her sooner, rather than just a few months before I left for the Navy. When I left in September, we both knew that something serious was afloat and we were going to stay in touch and see each other again.

Pensacola – Navy Basic Training

On 8 September 1950, my mother and brother dropped me off at Penn Station in Pittsburgh to catch a train for Pensacola. On the train, I met another cadet, Burke Eakle from West Virginia. At 26 years old, Burke, was at the top end of the age bracket for cadets. He had been in pre-medical school and decided that doctoring was not for him. He had always wanted to fly and he decided that it was now or never. We talked and, occasionally, sat quietly with our own thoughts on the rail trip south. I had much on my mind. Two years of college had matured me and I was looking forward to the challenge ahead. The last few hours of the journey from Mobile, Alabama, to Pensacola, Florida, were the toughest. We were on an old "Toonerville Trolley" train and the hot, humid air of the south made the last short leg of the trip seem endless.

When we disembarked in Pensacola, Burke and I found that we were assigned to class 20-50. This was the 20th pre-flight class of the year 1950. There were a total of 65 cadets assigned to the class and classes started every two weeks. That was almost 1,700 per year, but we were told that less than two thirds would make it all the way through to graduation. There were also commissioned officers from the military academies who had been assigned to flight training at airfields in the Pensacola area. As cadets, we would not get near an airplane for four months. This phase, called pre-flight, combined academics, physical training, and military instruction. It entailed a rigorous schedule, specifically designed to get us up to speed in the military way of life and prepare us for the upcoming flight training. It was "boot camp" for the naval aviation cadet.

The academic courses were similar to mid-level engineering and technical courses in college. The curriculum included instruction in basic aerodynamics, airframes, powerplants (engines), electrical systems, fuel systems, and hydraulic systems. We were also given a speed-reading course to help us study more efficiently. The most comprehensive course in pre-flight, air navigation, included celestial navigation and navigational trainers that simulated flying an actual mission.

The physical training was more rigorous than the academics. We took our academic courses in the morning and physical training in the afternoon. We worked out at a large gymnasium for three hours every day. This included calisthenics, wrestling, boxing, gymnastics, karate, and self-defense. Within about six weeks, everyone was in fine shape. I had lost most of my fat and gained a considerable amount of muscle.

We also received swimming and water survival training. We had swimming classes four days a week and the ultimate goal was to pass the Navy Triple-A swimming test. We swam long distances underwater to simulate swimming beneath water with burning gasoline on the surface. We also jumped from high towers into the water to simulate abandoning a sinking ship. Last, but not least, we rode the "Dilbert Dunker," an SNJ aircraft cockpit mounted on a steeply inclined track. The track ran from near the ceiling of the two-story pool building into the deep end of the pool. It was an experience I will never forget.

I entered the cockpit from a platform at the top of the track and was strapped in with parachute and flight gear. A horn sounded and the "Dilbert Dunker" streaked down the track at high speed and crashed into the water. It then sank to the bottom of the pool and turned upside down. At the end of this fabulous ride, I hung inverted, holding my breath with air bubbles and water turbulence around me. My heavy parachute pack, now waterlogged, was quite cumbersome. Conscious of my growing need for fresh oxygen, I released my seat belt and shoulder harness, placed my hands on the canopy bow, and pulled myself downward and away from the "Dunker" cockpit. With all my remaining strength, I swam toward the surface near the edge of the pool. This technique was important because if a cadet got excited and tried to go directly to the surface, he would strike his head on the rail superstructure that extended under the water.

The "Dunker" tested our physical and psychological abilities. Cadets who panicked had to be saved by Navy divers who were standing by. I made up my mind that I was strong enough to do it. I could swim well enough and it was just a matter of "keeping my head" and remaining calm. I thought of how my brother taught me to swim in the Ohio River years before and I decided that he would have made a good Navy swimming instructor.

In the evenings, several days a week, we had competitive sports with cadets from the other battalions. The instructors encouraged a high level of competition. I had played football in high school with "pads," but I was hit harder and bruised more playing Navy flag football. Everyone was in great physical condition and we literally "knocked hell" out of each other. It was all part of the Navy plan. They did not want any non-aggressive fighter pilots.

As part of our military training we wore uniforms, marched to and from all classes in formation, and saluted all officers. Marine drill instructors "put us through the mill" in marching and close order drill carrying the nine-pound M-1 rifle. The instructors gave demerits and penalties for any infraction of the rules, such as a shoe not shined properly or an imperfection of the uniform. Breaches in discipline were severely punished.

As my class progressed through pre-flight, some students were subject to attrition or "washing out". A few cadets dropped out because of academic difficulties. Some developed or discovered physical problems that disqualified them from flight training. Some were unable to cope with military discipline and others decided that it wasn't what they expected or wanted. We began pre-flight training with 65 cadets, but only 53 graduated.

Graduation from pre-flight was the first step to gaining the coveted Navy "wings of gold." I was in the best physical shape of my life. I had developed a great deal of confidence from having completed a tough preparation course for flying. As much as I looked forward to the upcoming courses, I also felt some anxiety about being able to handle flight training.

Naval Flight Training

In the early 1950s, Navy flight training consisted of two parts: basic and advanced. The basic stage was the same for all students while the advanced stage varied depending on what flying category the individual selected. Categories included fighter, dive-bomber, patrol-transport, and helicopter. The fighter and dive-bomber categories operated from aircraft carriers. The basic stage also included six carrier landings and qualification in the SNJ primary trainer. Thus, all Navy pilots had an opportunity to land aboard a carrier, even those who later went into transport aircraft and flew from land-based airfields. This was an excellent approach to training in that it gave all Navy pilots a look at what was involved in carrier operation. If there was something about it that they did not like, they had the opportunity to select the patrol-transport category when they went into advanced training.

Basic flight training took place on various small Navy airfields around the Pensacola area. The North American SNJ (Navy designation) or AT-6 (Air Force designation) "Texan" served as primary trainer. The SNJ was a low-wing, all-metal aircraft with two main landing wheels and a tail wheel. It held two pilots in a tandem cockpit. A Pratt & Whitney R-1340 radial engine that developed 550-horsepower powered the aircraft. An outstanding trainer, the SNJ was a demanding aircraft to fly. The tail dragger configuration limited

Courtesy of Don Mallick

Mallick climbing into the cockpit of a North American SNJ primary trainer.

forward visibility on the ground and the engine provided enough torque to keep the novice pilot busy trying to hold a steady course on takeoff.

In many endeavors, accomplishing a task involves both art and science. This is especially true of flying. The science portion includes knowledge of machinery, functions, numbers, limitations, and boundaries. The art involves coordination of eyes, ears, mental inputs, and physical inputs to fly the aircraft. A pilot must make control inputs based on all of the available cues and information and integrate them in his mind. When this is done properly, the pilot and aircraft almost become one and the aircraft's response is just an extension of the pilot. The science part is the easiest. An academic task, it merely requires study and memorization like other academic endeavors. The art is more difficult as it requires the student pilot to develop and hone his inborn responses and coordination to the point that he is the master of the aircraft.

Solo

No pilot ever forgets his solo flight. This was certainly true for me. It was the culmination of all my sweat and tears from learning to fly the SNJ. My training consisted of about 20 flight hours. Up to this time, I made all my flights with the instructor on board. Now I would fly alone for the first time, trusting my skill and training. My main instructor judged me qualified for solo flight, but before the Navy would let me go, another instructor had to provide a "second opinion." I was assigned a solo check instructor and I had to demonstrate my flying skills in all areas of flight including landings before he would release me to solo on my own. We took off from Whiting Field into a deep blue Florida sky dotted with scattered cumulus clouds. As we soared over the tall green pines that covered much of the countryside wind buffeted the open cockpit. Soon we arrived at Pace Field, a mile-square patch of light green grass surrounded by pines. Here I demonstrated my skill to the check instructor and he "kicked me out of the nest" to fly on my own for the first time. After he secured the rear seat and walked to the edge of the field, I taxied to the takeoff spot, went through my checklist, and pushed the throttle forward.

Without the instructor's weight in the back seat, the tail came up quickly and I was soon flying away from Mother Earth. The wind in the cockpit and the noise of the engine were all a symphony to me as I experienced the elation and thrill of being on my own and flying without any help or comment from the rear seat. Again I felt the buffet of the wind which caused my knee-board checklist to flop around a bit. The engine exhaust crackled as the pistons and valves provided power to turn the propeller. As Pace Field dropped behind my left wing, I knew that I had reached the 180-degree position and it was time to start my descent and turn for landing. I pulled the throttle back and the crackling exhaust quieted slightly. The aircraft responded to my pitch and roll inputs and I began my turn toward final approach. I rolled the wing gently to straighten my approach and verified that I was at the correct speed and altitude.

Crossing the runway threshold, I chopped the throttle and pulled back slightly on the control stick to make my landing flare. Just as I had been trained to do, I stalled the wing just at the moment of touchdown and completed a successful landing. I repeated my performance twice more with great precision. After the third landing, I taxied back to pick up my check instructor who had been "sweating it out" from beneath a tree on the edge of the field. I took off and flew back to Whiting Field.

I had completed Stage "A" of my training, basic flight and solo. After that, I continued with increasingly more advanced training. Stages "B" through "G" consisted of Precision Flying, Acrobatics, Instrument Flying, Night Flying, Formation Flight, and Gunnery Practice. I had to complete them all to become a naval aviator.

Stage CQ, Carrier Qualification

Next came the greatest challenge: Carrier Qualification.

So far, our Navy training syllabus had been fairly similar to Air Force training. The next stage, called "CQ," required cadets to gain proficiency in flight operations aboard aircraft carriers. It was the stage that some said "separated the men from the boys." I had accumulated 185 flying hours in the SNJ. I knew the aircraft well and felt very comfortable flying it. Still, I wondered if I could land it on the deck of an aircraft carrier as it pitched and tossed on the ocean. The experience involves all of the same techniques applied in land-based flight operations except that the runway is now floating (and moving!) in the middle of the sea.

My first experience with carrier-type landing approaches took place at an airfield on dry land. The runway was marked on the ground to simulate an aircraft carrier landing deck. Here I practiced my technique on a stationary target. Airfield carrier practice flying was exhilarating. I spent many hours flying approaches to the simulated landing deck. The practice areas were small, remotely located fields carved into the Florida pine forests. They were triangular in shape so a pilot could always select an approach into the wind.

During my first simulated carrier practice, I sat in the rear seat while the instructor made the landing. The aircraft was in a "dirty" configuration, meaning that landing gear and flaps were extended. We made our approach with about 68 knots indicated airspeed, 17 knots slower than the normal approach speed for a land-based runway. We flew so low that it appeared you could reach out and touch the tops of the nearby trees. The instructor pilot described what he was doing as he flew the approach but he also watched me in his rear view mirror. He must have seen me focused on watching the airspeed indicator. "Damn it, Mallick," he exclaimed, "quit watching that airspeed indicator!" And with that he stamped hard one rudder pedal. To my surprise the aircraft did not stall and drop into the trees as I might have expected. The instructor

Courtesy of Don Mallick

The North American SNJ "Texan" served as the primary trainer for Naval Aviators in the Basic Phase. This photo was taken during carrier qualification training in September 1951. The tail-dragger configuration and powerful engine torque made the SNJ a demanding aircraft to fly.

flew about six patterns and approaches and I became a little more comfortable each time. It was a new realm of flying, but I quickly became accustomed to it.

On the next flight, I sat in the front seat and my instructor watched from the ground. The instructor couldn't do much from the back seat anyway. A Landing Signal Officer (LSO) played one of the most important roles in this exercise. The LSO was a qualified aviator trained in the art of directing pilots as they made their carrier landings. The LSO spotted deviations from a normal approach even before the pilot did. He used two large paddles, one in each hand, to provide visual signals to the approaching pilot. These signals told the pilot what corrections to make to bring his approach back into alignment. These signals included: Too Fast, Too Slow, Increase Turn, Decrease Turn, Too High, Too Low, Skidding, and combinations thereof. Two other important signals were Cut Power and the Wave Off (or Go Around). The first set of signals were suggestions, the last two were orders. Cooperation between the pilot and LSO was essential for orchestrating a good carrier approach.

I usually flew with the canopy open because the Florida temperatures were pleasant, even in October. After a few flights, I became comfortable having the tall pine trees close below while my airspeed hovered around 65 knots. I pushed thoughts of engine failure and crashing into the trees to the back of my mind and concentrated on flying the approach. With wind in the cockpit, engine noise, and crackling exhaust, it was as close to true "seat of the pants" flying as one comes. I enjoyed it immensely. Even the afternoon turbulence that bounced the aircraft around could not erase the thrill of flying a good approach. My pulse raced as I watched for the Cut Power signal, chopped the throttle, and flared to a spot landing. Once again, I pushed forward on the throttle and my plane roared down the runway. Soon, the main gear lifted off the ground followed closely by the tail wheel. I pulled back on the stick, just clearing the trees at the end of the field, and re-entered the pattern for a repeat performance.

Field carrier practice lasted two weeks. I accumulated 11 flight hours and 53 field carrier landings during 15 sorties. My instructors soon deemed me qualified to land on the U.S.S. *Monterey* (CVL-26), a small carrier stationed at Pensacola and used for carrier qualification training. My instructors assured me that landing on the *Monterey* would be easier than the field carrier practice.

On my carrier qualification day, 16 October 1951, I was a "walk-aboard." A number of cadets actually walked aboard the carrier at the Pensacola Pier before it sailed. The other cadets flew out in SNJs to join the pattern once the carrier was cruising in the Gulf of Mexico. When one of these cadets completed his required six landings, he exited the aircraft on the flight deck and one of the walk-aboards climbed into the cockpit. As I waited my turn in the ready room, the U.S.S. *Monterey* wallowed in the swells of the Gulf. The ship vibrated and shook as the engines turned the screws, pushing the ship through the water to generate an adequate wind over the deck for the flight operations. The ready room was small, with limited air circulation, and I began to feel a little nausea. This concerned me and I wondered what kind of a Navy pilot I was going to make if I was susceptible to seasickness. I heard my name called over the ship's intercom and I made my way to the flight deck. The cool sea breeze revived me and I felt much better. Soon, I was strapped into the cockpit and prepared for takeoff. As my plane lifted off the deck, the airflow through the cockpit increased and I felt myself fully restored. I completed six good approaches and landings with no wave-offs or difficulties. As aviators like to say, it was "uneventful."

The instructors were right. It was easier than the field carrier training. For one thing, I didn't have to look at all those pine trees. Because I had been a walk-aboard, I flew the aircraft back to the mainland. I can't ever recall being more pleased with my performance than I was this day. The seasickness was pushed to the back of my mind and never troubled me again.

Advanced Flight Training

When I finished basic flight training, I had approximately 200 hours flying time and I had accomplished all of the phases that would normally have been required for combat. The move to advanced training allowed me to experience flight in larger aircraft with higher performance. Otherwise, the training syllabus was generally similar to basic. During basic, I observed that there were some "natural pilots" among the cadets, those who easily and quickly adapted to

Smell of Kerosene

Courtesy of Don Mallick

Produced in vast numbers during World War II, the Grumman F6F Hellcat was well suited to carrier operations. Mallick (shown here around 1951) achieved advanced carrier qualification in the Hellcat just prior to receiving his commission.

flying. I would estimate that about five percent of our class fell into this category. Unfortunately, I was not one of the natural pilots. I was able to retain new capabilities with precision and consistency, but it took me longer to master my tasks.

In mid October 1951, I reported to advanced flight training in Corpus Christi, Texas. Corpus Christi was a lovely, clean town. It seemed hardly large enough to be called a city, but it had a "homey" atmosphere and very friendly people. There were several training bases in the area, including Mainside and Cabiness Field. Kingsville, about 40 miles from Corpus Christi, was used for transition training to jets, something that I looked forward to. Mainside provided multi-engine training and the advanced instrument flight school. I was assigned to Cabiness for advanced training for fighter and attack pilots. There were three fighter squadrons flying the F6F Hellcat and one squadron flying the high-performance F8F Bearcat. An attack squadron flew the AD Skyraider. I was assigned to an F6F Hellcat squadron.

The Hellcat had much higher performance than the SNJ. It was equipped with a 2,000-horsepower engine and could attain speeds just over 400mph. By comparison, the SNJ had a 550-horsepower engine and flew about 200mph. Operating the F6F aircraft was a greater challenge than flying the SNJ. I performed the same kinds of tasks (formation, gunnery, and acrobatics), but the demands of the high performance aircraft were greater. In advanced training, I flew single seat fighters. They had no back seat for an instructor. Usually two instructors flew "chase" in accompanying aircraft. One served as lead and another flew in trail to keep a close eye on the students.

Unfortunately, we experienced very high losses in the first few months of advanced training. It seemed like a terrible streak of bad luck. Between the middle of October and the Christmas break, we lost five cadet pilots to accidents. Four died as the result of mid-air collisions. Another crashed after suffering from carbon monoxide poisoning in his cockpit. The tragedies weighed heavily on the cadets in my class, but we continued our training until the regularly scheduled break.

I was glad to get home for Christmas. Audrey and I had announced our engagement in the summer of 1951 when she came to visit me in Pensacola. We were going to be married as soon as I received my wings and commission. I needed the break to catch my breath. Now that I was starting a family, my realization of the risks of flying really hit home. Those thoughts, previously in the back of my mind, were now brought to the forefront. The vacation break helped me get my bearings.

After the holidays, I returned to training with a refreshed body and mind. I renewed my studies with vigor. Training progressed well for a month or more and then we had another accident. Jack Cliff from Hershey, Pennsylvania was one of our best natural aviators. He was tall, athletic, handsome, and he excelled in academics and flight all through the program. Jack led his class through all stages of the training program. He was assigned to the F8F Bearcat squadron and was the top cadet in that group. I admired Jack as a person and an aviator. Perhaps, if I ever wanted to be someone else, it would have been Jack.

He completed the advanced syllabus and had his orders to report to Pensacola for advanced carrier qualification in the F8F prior to receiving his wings and commission. As he was cleaning out his flight locker, an instructor came by and asked him if he would like to fly "safety chase" for the instructor during an instrument flight. The safety chase provided a visual lookout while the instrument pilot concentrated on "hooded" flying. Jack jumped at the chance for another flight and suited up. When the instrument portion of the mission was completed, Jack led the flight of two Bearcats back to the landing field. The traffic pattern was at an altitude of 800 feet with a sharp 180-degree break for the downwind turn. In a normal fighter break, the pilot executes a near 90 degree bank turn and pulled three to four g's as the aircraft slowed to landing speed. When Jack made the break, he pulled smartly into the turn only to experience a high-speed stall and he rolled into a spin and crashed. The aircraft hit so hard that the 20mm cannons were embedded eighteen feet into the ground. Jack was killed instantly.

I was stunned. The other students and instructors shared my grief and confusion. We couldn't understand how this could happen to someone as sharp as Jack. The general consensus was that Jack's exuberance over having just completed advanced training and an exhilarating flight led him to pull too many g's at the break point. How could someone as talented as Jack let this happen? A pilot must know his capabilities and live within them. Unfortunately, over years of flying, I have observed other similar accidents. An Air Force flight surgeon involved in accident investigations made the following observation concerning pilots: "The main problem is overconfidence. Pilots have difficulty saying, 'I can't.'" Excessive motivation sometimes pushes pilots beyond their limitations.

About three weeks after Jack's death, I completed advanced training and headed to Pensacola for advanced carrier qualification. I felt elated, knowing that this next phase was the final step to receiving my wings. I was also looking forward to graduation because it meant that Audrey and I would soon be married. By this time, we had known each other for over two years and were anxious to start a family.

Advanced Carrier Qualification

The preparation for advanced carrier qualification began with field carrier practice in the F6F. I made 82 practice field carrier landings before attempting to operate from an actual aircraft carrier. Once again, the U.S.S. *Monterey* served as the training carrier and it hadn't grown an inch larger since basic. Fortunately, the F6F was well suited to carrier operations. When I cut power to land, the airplane settled in for a solid landing with no tendency to bounce or float. I was required to make 12 landings to qualify. I completed seven landings my first day out and five the next day. I had no trouble qualifying. In fact, I earned my highest flying grades during this phase of training. I completed it in 21 days with 21 flying hours, including the 12 actual carrier landings.

I qualified on 28 February 1952 and received my commission and wings on 5 March. Audrey waited patiently and we celebrated our wedding a few days later in Pittsburgh. We then headed to Corpus Christi, Texas, for my advanced instrument training.

Instrument Training

I had to complete two advanced training schools, Instruments and Jet Transition, before assignment to a fighter squadron. Instrument school served an important purpose in preparation for Jet Transition. It involved flying the twin-engine propeller-driven Beech SNB-2 Navigator. I studied hard, and the training paid off throughout my flying career.

In a little over a month, I had flown 42 hours of instrument and night flying. The SNB was a two-seat aircraft, but the instrument flights were structured with the trainee pilot flying the aircraft and managing all the systems just as he would do in a single-seat aircraft. The SNB had four fuel tanks and the student pilot was responsible for managing the fuel and switching the tanks when they reached 20 percent capacity. If the student forgot, the instructor would allow the fuel to reach 10 percent before switching the tanks for the student. If the student failed to properly manage his fuel consumption, he received a demerit or bad grade. Thanks to my highly qualified instructor, I excelled in the instrument school and enjoyed the flying.

Next, I was assigned to Kingsville, out in the country, about forty miles from Corpus Christi, for jet transition. I had applied myself to instrument training with the idea that jet training would be my reward. The jet school used both the Lockheed TO-1 (Navy version of the F-80 Shooting Star) and TO-2 (Navy version of the T-33, a two-seat F-80). The TO-1 was a beautiful little jet with smooth lines and a small bubble canopy. There was a hand crank to close the canopy and the pilot had to duck to allow the metal canopy bow to pass over his head. Once it was closed, visibility was pretty good. Everything in the cockpit was within easy reach. The airplane had an ejection seat escape system, something new, associated with jets. The high speed of the jet required a special seat to help the pilot clear the aircraft during a bailout. Early models of the F-80 family didn't have ejection seats and the pilot had to extract himself from the cockpit in the event of an emergency. It was very dangerous. The new seats increased the survivability of jet accidents.

The Navy instructed us in the theory of jet propulsion and operation of gas turbine engines. We learned that they really gobbled fuel at low altitude and that the optimum realm of operation was at high altitude (around 30,000 feet). We learned to monitor tailpipe temperature or TPT (today known as Exhaust Gas Temperature or EGT), a critical item. Jet engines, introduced in the early 1940s were still relatively new and prone to operational problems. The instrument panel included numerous warning lights. It was a world apart from the old F6F. When we initiated the self-test position on the warning lights, the cockpit lit up like a Christmas tree. Early jet engines also suffered from very slow acceleration. It took perhaps 8 to 10 seconds for the engine to go from idle to 100 percent power. A pilot who was accustomed to almost instantaneous power of a reciprocating engine had to be very cautious and anticipate his requirements for power in the jet.

I noticed that jet aircraft seemed smoother to fly and quieter than propeller aircraft. Gone were the heavy vibrations of the reciprocating engine. In their place was a uniform, high-frequency hum. The speed seemed to be several magnitudes higher. The aircraft itself had clean lines and low drag. In fact, jet airplanes needed speed brakes to enable the pilot to decelerate in combat or during landing.

The instructors at the Jet Training Unit (JTU) were some of the Navy's most experienced aviators. They were hand-picked to transition new pilots who had been trained only in propeller aircraft. One of these instructor pilots, Lt. Cdr. Dean "Diz" Laird, was a joy to behold while flying in air-to-air engagements. Normally in formation work, the leader does not "carry full power" on his aircraft. This allows the wingman a little margin on power to help hold his position relative to the leader. Diz, however, believed a pilot should fly "balls out" or full-power if he wanted to survive in air combat. The students flew with as many of the instructors as possible to gain a cross section of their experience. Typically, a student flew as the instructor's wingman during

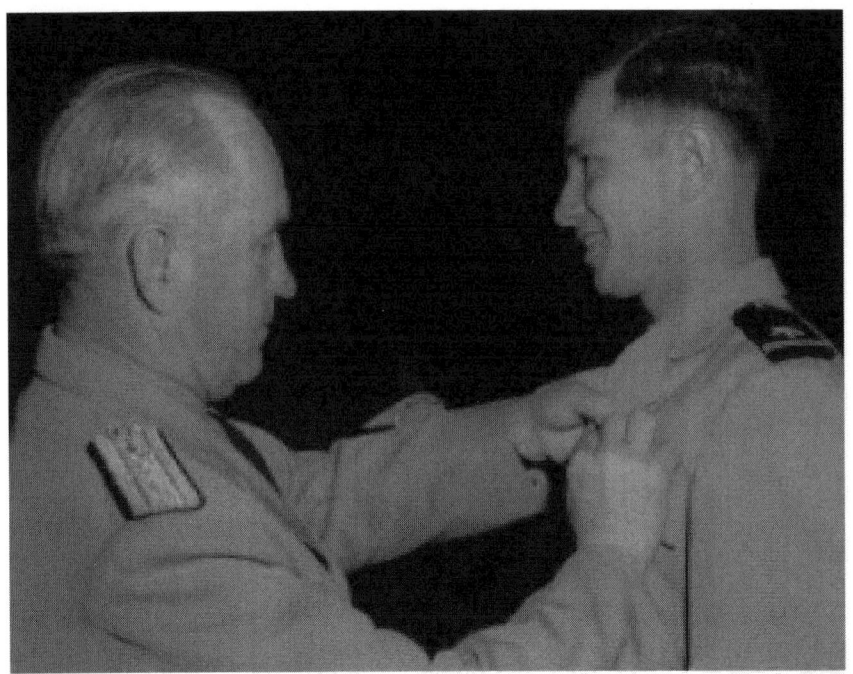

U.S. Navy photo NA933882

Mallick receives the coveted "wings of gold" at graduation from naval training on 5 March 1952.

air-to-air training engagements. The students called Diz "100 percent Laird" because he always flew at full-power and it was difficult to stay with him. The only way was to turn inside him during maneuvers. My flight with Diz was a memorable one.

Diz and I were flying against two other TO-1s, flown by another instructor and a student wingman. We picked up the other section (two-ship formation) visually and it was a head-on engagement. A couple of level tactical turns resulted in neither section of aircraft having the advantage. About this time, Diz took us vertical. We went straight up, trading our speed energy for altitude. The idea is to come over the top and spot your enemy down below and have the advantage of altitude and position in order to set up a gunnery run on them. Diz had timed his recovery at the top perfectly with just enough speed to maintain flight. I was at 100 percent power, turning inside him to keep him in sight. As I reached the top of the wing-over, my aircraft stalled and entered a spin. The maneuver was very disorientating and I struggled to recover from the spin and continue flying. I lost some altitude, but managed to recover after three turns. As I pulled up to rejoin the mock combat, I wondered if Diz had seen me spin. I didn't plan to mention it myself. "By the way Mallick," Diz said later as we debriefed the mission, "not a bad spin recovery, I was glad to see you get back into the action!"

The instructors at JTU were highly dedicated to teaching each of us as much as possible and improving our piloting skills. I found the JTU to be all that I had expected and more. It was worth the wait, plus all the work I did to get there. Following my first flight in the TO-1, I walked toward base operations feeling as if I was on "Cloud 9." Several other jets were returning to the parking ramp. As they taxied by, the odor of the exhaust fumes took my mind back to earlier years on my aunt's farm with kerosene lamps. The kerosene-based jet fuel had nearly the same mix as used in the lamps. Since then, I have always equated the smell of kerosene with my life as an aviator.

2

Naval Air Operations

Fleet Operations

JTU went by almost too quickly and it was time to move on. I had orders to report to Naval Air Operations, Atlantic Fleet (known in Navy parlance as AIRLANT) in Norfolk, Virginia. The assignment officer tried to place me in a night fighter squadron in Key West, Florida. I did not particularly want night fighters as I had heard that night fighter squadrons were heavy with "brass" (higher-ranking officers) and they were doing all the jet flying in the limited number of jets available. Consequently, a junior pilot like myself faced another year flying F4U-5N and F6F-5N propeller-driven night fighters before getting into jets. This was not the case in the day fighter squadrons, all of which were currently equipped with jets. I wanted to fly jets.

When I told the assignment officer that I would sooner turn in my wings than fly night fighters, he was shocked. I demanded assignment to a day fighter squadron anywhere on the East Coast. It was really just a bluff. I would never have turned in my wings after all the work I had invested to get them.

Two months later, Audrey and I left for Norfolk to pick up my final squadron assignment. It was a pleasant day and Audrey waited for me in the car when I entered the AIRLANT building. I said a few prayers on the way to the assignment desk. Maybe someone was listening. I was assigned to a day fighter squadron called VF-172 in Jacksonville, Florida. Elation washed over me. My bluff had worked.

Navy Jacksonville, VF-172

Fighter Squadron VF-172 had a very colorful reputation. The squadron name "The Blue Bolts" was represented on each aircraft by long slashing lightning bolts painted on the tip tanks. The squadron had recently returned from a combat tour in Korea, five months before I reported for duty. The unit had a good combat record and came through the hostilities with no losses. The squadron's heroics were later immortalized by author James A. Michener in *The Bridges at Toko-Ri*. To the chagrin of veteran VF-172 pilots, a movie adaptation of the story featured F9F-5 Panther jets instead of the F2H-2 Banshee they had flown. When I reported in to the squadron, the commanding officer (CO) was Commander James B. Cain. A dashing, handsome aviator, Jim Cain was exactly what one would expect for the CO of a Navy fighter squadron. He could have played the lead in the movie and as it turned out, he was one of the technical advisors.

New Navy pilots at the time were assigned to sea duty for three years and then shore duty for one year. Each unit had a land base, such as Jacksonville, for training when it was not at sea, but the squadron could be assigned to a carrier at any time during its three-year sea duty assignment. Shore duty assignments included instructor's slots in the training command. Flying assignments were designed to provide a cross-section of pilot experience in the squadron. Fleet squadrons checked out new pilots in the aircraft and qualified them on the carrier. The new pilots were brought up to speed as rapidly as possible, especially if the squadron was scheduled for an upcoming cruise.

When I reported in, I was pleased to find several of my good friends from flight training. Burke Eakle, the fellow who had left Pittsburgh with me many months before on our way to Pensacola; Don Minnegeroede, one of my pre-flight class roommates; and Jim Osborne, a cadet from the pre-flight class behind mine were all there. Jim had caught up and followed me through Training Command. The presence of these friends made the transition to squadron life more pleasant. Together, we faced the same checkout and training requirements as we prepared for carrier duty in the F2H Banshee. Another Training Command friend, Desmond Kid, showed up about four months later. Des graduated about the same time as the three of us, but the Navy assigned him to helicopters against his wishes. Des had been dating a girl whose father was a Navy Captain assigned to the Pentagon. Through this contact and many letters, Des was finally reassigned to a fighter squadron.

Flying the Banshee

The F2H-2 Banshee entered production in 1949 and proved itself in Korea. The F2H series was the second Navy aircraft built by McDonnell Aircraft Company in St Louis, Missouri. Their first was the FH-1 Phantom which served briefly until the advent of aircraft with better performance. The F2H-2 had a number of improvements, including greater range and endurance. The Banshee was the electric airplane of its time. Almost all of the actuators were electric rather than the hydraulic actuators found in most contemporary aircraft. The Banshee had electric landing gear, electric speed brakes, electric flaps, and electric starters on the J34-WE-34 jet engines. Even the aileron boost was an electric-hydraulic pack type system. The wing-fold system was electric. The rudder and elevator were manual with aerodynamic boost tabs to assist the pilot. The aircraft had a large 200-gallon tip tank on each wing, for added range. This also increased the takeoff roll on a warm day. The aircraft featured a conventional straight-wing design with large deep wing roots that housed the engines, one on each side of the fuselage. The overall appearance was sleek for its day.

The flying qualities of the Banshee were excellent. It was a stable aircraft, easy to fly on instruments and yet quite maneuverable in air-to-air engagements.

Courtesy of Don Mallick

Crews prepare McDonnell F2H Banshee aircraft for sorties from the U.S.S. Franklin D. Roosevelt (CVB-42). The wings folded so that the airplanes took up less room on the crowded flight deck.

One of the biggest drawbacks of the Banshee was that it was somewhat underpowered. This was typical for most contemporary jet aircraft. Another problem developed with operational use. As the Banshee accrued more flight hours and years of service, there were several wing failures during maneuvering flight. To compensate, McDonnell designers added flight control system modifications. They added a heavy bob-weight device in the pitch axis controls to help the pilots avoid over stressing the aircraft. Additionally, the Navy reduced flight envelope "g-limitations" for the last few years of the Banshee's life. Multiples of the equivalent of Earth's gravitational force, generated during maneuvering flight, are called "g-forces." Aircraft structures are limited to specific maximum g-forces that they can sustain without damage. Armament on the Banshee consisted of four 20mm cannons in the nose and rocket and bomb rails. This allowed the Banshee to serve as a fighter-bomber, its primary role in Korea.

My checkout and familiarization in the Banshee went quickly. It was easier and more comfortable to fly than the TO-1. The Banshee was also much more stable to fly on instruments. I flew my first checkout flight in mid-June 1952 and I had 30 flight hours in the aircraft prior to my first carrier landings. On 8 August, I earned my carrier qualification with the nine required landings aboard the U.S.S. *Franklin D. Roosevelt* (CVB-42).

First North Atlantic Cruise (September 1952)

The Navy maintained a presence in the Mediterranean Sea and there seemed to be at least one aircraft carrier there at all times, sometimes more than one if war threatened that part of the world. Aircraft carriers came in two sizes. The *Essex* Class, the largest available in World War II, was the smaller of the two. Of the 24 built, the first had a length of 855 feet and the remaining examples were 888 feet long. The larger *Midway* Class consisted of three examples: *Midway* (CVB-41), *Franklin D. Roosevelt* (CVB-42), and *Coral Sea* (CVB-43). They stretched 968 feet from bow to stern.

All of these ships were straight-deck carriers. Operating jet aircraft on these ships was challenging and there was little margin for error. The jets were operating on flight decks designed for F6Fs, F4Us, and AD Skyraider type airplanes, but they had a higher approach speed. Bolters ("go-arounds") were unavailable to us on the straight deck. If you landed, you landed, good or bad. The angled flight deck, installed on the U.S.S. *Midway* in 1957 (and still in use on modern carriers), helps a great deal in jet operations. It reduces the required amount of arresting gear by half and moves aircraft takeoffs and landings away from other airplanes parked on the deck. Carrier flying is still tough demanding flying, but the steam catapult and angled deck have improved jet carrier operations immensely.

Our carrier qualification landings had been made just off the coast of Florida, near Jacksonville. The weather had been pleasant and the ocean was relatively calm. On our way to the Mediterranean, we lost the nice weather as we crossed the Atlantic Ocean. The weather in the North Atlantic in the fall was usually bad. We experienced rain, wind, violent storms, and rough seas. The angry ocean dwarfed the giant steel hull of the aircraft carrier. The decks rolled and pitched in the towering waves, with the fantail sometimes rising or falling more than 20 feet.

I always found flying in this weather stressful. After a descent through thousands of feet of clouds and rain, I might break out with a 400-foot ceiling (if I was lucky) beside the carrier. Proceeding ahead of the carrier, my division would break in 45-second intervals and turn to a downwind leg in preparation for landing on the carrier. The ocean, deep green beneath the steely overcast, was covered with whitecaps whipped by the wind. The ship's wake, normally a handy reference to find the carrier again when I looked up from my cockpit scan, had disappeared as it was churned and dissipated by the waves. I now had two options. I could make a landing on that pitching and rolling carrier or take a dip in the now ugly ocean. Like most Naval aviators, I selected the first option and did my best.

I pushed worries about the weather, the angry sea, and even the pitching flight deck of the carrier into the back of my mind. I concentrated on my altitude control, using the bridge "island" on the carrier as a reference, and watched for the signals of the LSO. As I approached the ship, the flight deck

Courtesy of Don Mallick

Observers watch as McDonnell F2H Banshee jets line up for takeoff from the deck of the U.S.S. Franklin D. Roosevelt.

rose and fell in front of me. Concentrating on it now would be a mistake, causing me to start "chasing" the deck and fouling up my approach. The LSO giving me my landing signals watched the flight deck for me. If the deck seemed reasonably close to a good position, he gave me the cut signal. In that case, I pulled the throttle back and landed. In the end it was my judgement call. If I couldn't make it, then I was out of luck. I only had one chance. Either I made it aboard in some manner or I would splash into the churning sea.

There were many occasions operating in this rough weather when the LSO waved me off and sent me around for another approach because the flight deck was pitching too much. Even with a good approach, sometimes a pilot couldn't achieve synchronization with the heaving deck and could not get aboard. Often, he tried anyway. This type of flight operation was risky and we broke some airplanes with hard landings. I give the Navy credit for acknowledging weather as a major factor when the accident reports were made. The Navy knew from prior experience that there were going to be such accidents during training, but they apparently considered it an acceptable risk.

As a relatively new pilot, I had logged fewer than a dozen carrier landings in good weather. Weather conditions were so bad that even our more experienced pilots were having accidents. Consequently, I felt some trepidation when I taxied onto the catapult and waited until the bow of the ship had

Naval Air Operations

bottomed out in its pitch cycle. As the deck started upward, the catapult officer gave the signal to fire and my jet launched from the carrier when the bow pointed above the horizon. Even with these precautions, I sometimes went through a large curtain of water spray as it exploded up from the bow of the ship. As I retracted the landing gear and climbed away from the carrier on my mission, I realized that was the easy part. The hard part, getting back aboard ship, was yet to come.

Soon, I had two or three landings under my belt in the infamous North Atlantic. One day, I was returning to the carrier as Number Two man in a flight of four. We normally operated with a division of four aircraft. My division leader was Lt. Bob Hamilton, nicknamed "Hambone." All Navy pilots had a nickname. Mine was "Jake" because the current Russian delegate to the United Nations was Jacob Malik.

As we approached the carrier, I made my break behind Hambone and entered the downwind leg of the approach. It was hard not to look at the huge waves and streaks of salt spray streaming off the top of the whitecaps. When I reached the 45-degree position, I noticed that Hambone had landed and was clearing the arresting gear area. I watched the LSO and started to work the approach. As the carrier loomed in front of me, I felt uncomfortable with the approach. I watched the pitching deck and received a lot of correction signals from the LSO. I took a wave-off and went around just short of the normal cut position. On the second approach, I felt much better. I wasn't distracted by the pitching deck and the LSO didn't give me as many corrections. I landed in good shape.

After the recovery of all aircraft, various LSOs visited the ready rooms and debriefed the pilots on their approaches. It was an excellent learning and feedback system. There were times that the LSO observed things that the pilot would never had been aware of. "Mallick," one LSO asked me, "why did you wave off on the first pass?" I told him that I just felt I was behind the approach and was not comfortable enough to come aboard. He acknowledged that with no further comment. After the debriefing, Hambone took me aside and explained the importance of getting aboard quickly. Of course, that was basic knowledge. All carrier pilots knew this. I didn't pay much attention to Hambone because I figured a few minutes extra wasn't worth busting up an airplane, or me for that matter. The nice thing about flying a single-pilot fighter was that I was in charge and no one else. The next day, Hambone made a hard touchdown and broke the landing gear off his airplane. His accident was blamed on the pitching deck.

I had developed the habit of observing flight operations from the bridge or "Pri-Fly" when I wasn't scheduled to fly myself. This gave me a chance to observe and learn from watching other pilots and the LSOs as the planes came aboard. It was interesting to see the pilots respond to the LSO signals. Later, after I was assigned as the squadron Structures Officer, I watched the landings with another motive. If I observed a hard or abnormal landing, I made a note of it and had my maintenance chief make a special inspection of the aircraft involved. I didn't want to send another pilot out with a damaged aircraft.

Fortunately, the North Atlantic operation did not last long and we finally headed through the Strait of Gibraltar and into the beautiful Mediterranean. The "Med" was a critical strategic body of water with many countries on or near its shores that influenced international affairs. The presence of U.S. Navy ships and aircraft in the region was a vital part of our foreign policy and remains so today. Flying off the carrier in the Mediterranean was a welcome relief. There were occasional storms to contend with but I never saw the tremendous swells, waves, and wind that caused the flight deck to heave and roll like it did in the North Atlantic.

On one flight, I experienced a hook-point failure during landing. The approach had been normal with a cut, flare, and reasonable touchdown on the carrier deck, but the rapid deceleration of the hook picking up a wire never occurred. I felt a little slow down and then a release and the aircraft seemed to accelerate toward the barriers. I felt a couple more grabs on the aircraft but it didn't stop. I reached forward and verified that my hook handle was down. As I engaged the first Davis Barrier, it snagged only my left main landing gear and the Banshee started to swing toward the side of the ship. For a moment, I thought that I was on my way over the side. At the last instant, my right wingtip tank caught in the vertical webbing of the large barricade and the aircraft came to a stop. I sat facing out over the left side of the ship and breathed a sigh of relief. Within seconds the deck crews had chocked the aircraft and I shut the engines down.

Initially, when the aircraft failed to stop, I thought that I had screwed up and forgotten to lower my hook. Fortunately, that wasn't the case. As I touched down on the steel flight deck, the hook point had broken in two, allowing a little more than one half the hook point to remain on the hook arm. What remained of the hook had a sharp edge that cut every cross-deck wire that it picked up, only slowing the aircraft slightly as it briefly caught each one. The final damage summary indicated that three cross-deck wires were cut completely in half and two others were kinked and had to be replaced. It took the deck crews more than twenty minutes to make the flight deck ready to recover more aircraft. In fact, the ship's captain passed the word to the ready room on the "squawk box" (P.A. system) to congratulate me. My aircraft had fouled the deck for 22 minutes, a record on the U.S.S. *Roosevelt*. The Navy always strove for a ready deck with no delays to airborne aircraft. Usually the flight deck was back in operation within a matter of minutes after an accident. Mine was not a record to be proud of, but you can't believe how happy I was to learn that it had been a materiel failure and not a pilot error.

This barrier on the deck of the Roosevelt *saved Mallick from a dip in the Atlantic Ocean on 28 August 1952.*

Some of the other squadrons suffered fatal accidents. One that occurred during daylight operations involved an Air Force exchange pilot who was flying the F2H-2 in our sister squadron, VF-171. The accident occurred in the North Atlantic during very rough seas and bad weather. I had not been assigned to fly that day so I was observing the recovery of aircraft from the Pri-Fly walkway. The pilot approached his landing position and received a cut signal from the LSO. For some reason, after the pilot cut his engines he decided not to land and he attempted to go around. As the pilot tried to bring the jet engines back up to speed, the Banshee settled toward the flight deck. The pilot rolled into a left bank to try and clear the carrier, but he struck a barrier stanchion. The aircraft cartwheeled over the side and into the ocean. A rescue helicopter and a destroyer searched for the pilot but he was never found.

Another daylight operational accident involved a McDonnell F2H-3, sometimes called the "Big Banshee." The aircraft and pilot were from a special detachment of a Composite (or VC) squadron. The F2H-3 had the capability of delivering a nuclear bomb (usually referred to as a "special weapon"). Security onboard the ship increased whenever a simulated or practice nuclear flight operation took place. One of the modifications allowing the F2H-3 to carry the "special weapon" was a modified pressurization system on the main landing gear struts. This allowed the struts to be pressurized above normal levels, causing the aircraft to sit higher above the deck for more clearance. This extra pressure was bled off prior to landing to place the aircraft landing gear system back to a normal configuration.

The VC detachment flew frequently for proficiency and this accident occurred while landing aboard the carrier. The pilot completed his landing checklist properly but, unknown to him, his gear strut depressurization system failed to function. His landing gear struts were extended more than normal and the nitrogen pressure in them was very high. He made a firm touchdown, but the airplane bounced and missed the arresting wires. A 20-foot-high jet barricade snagged the Banshee out of the air and slammed it to the deck. Coming to a stop, the Banshee struck several aircraft parked forward but, fortunately, there was no fire. The very large bundle of nylon straps that make up the top of the barricade slid up the rounded nose of the aircraft and dropped into the cockpit and pinned the pilot to his seat. A large part of the force stopping the aircraft was applied to the pilot's body. He was crushed and killed instantly.

Night Carrier Qualification

By the time we returned to Jacksonville, I had accumulated 125 hours of Banshee flying and 44 carrier landings. I was pleased with my performance on the cruise and very happy that I hadn't bent any airplanes. Fortunately, my squadron had only suffered a few minor accidents and had not lost any pilots.

I was now ready for night carrier qualification. As ready room talk turned to this new challenge, the pilots' wives naturally became aware that we had taken on a new task with an increased level of risk and danger. Audrey was certainly worried about it. She had become pregnant in January, a month after I returned from the Mediterranean. We were both pleased about it and looked forward to starting a family. The fact that she was pregnant probably caused Audrey to worry more than usual. I tried to shield her as best I could by not talking about the risk or making a big deal about it. I told her that I could not worry about it myself because that could affect my performance. I hoped that she could do the same.

Night field carrier training took place at NAS Jacksonville. It soon became apparent that this was not the ideal place. There were so many other ground lights in the area that it was difficult to make out the LSO or the landing spot marked on one of the runways. The LSOs wore flight suits trimmed with a fluorescent material, but they were still difficult to see. As I approached the 45-degree position, the LSO appeared as a small bright dot that slowly grew and then sprouted arms, at which time I could see his signal. On my first pass at Jacksonville, I knew where to look for the marked spot on the runway. I also knew where the LSO should have been, but with the clutter of background lights in the area I never saw him. I flew the pass on my own, knowing where

I should touch down. I cut power and made the touch a[...] back into the pattern, the LSO came on the radio to comm[...] "Not bad, Mallick," he said, "but you were a little high al[...] to myself and answered him, "Hell, if you think that wa[...] actually see you on the next one." At the debriefing foll[...] session, the other pilots confessed that they had similar prob[...] field carrier practice moved to Cecil Field which was loc[...] from city lights. This provided a better simulation of the task because there were very few lights on the carrier and the surrounding sea looks black at night.

On the afternoon of 9 February 1953, four of us flew out to the U.S.S. *Coral Sea* (CVA-43) which was on station about 150 miles east of Mayport, Florida. We made two daylight carrier landings as a refresher and preparation for night qualification. After dark, two of the Banshee pilots launched while the rest of us waited for the second mission. When the first pilot completed his second landing, he parked on the port catapult and left the engines running. I crawled into the aircraft, strapped in, and went through my checklist. I noted that the fuel remaining was 2,500 pounds, sufficient for two carrier landings but not enough to get to the beach if a problem occurred that would require diversion to a land field. I decided that it was too late to ask for more gas. I expected to do some good night flying and qualify, not goof around. I signaled that I was ready and set the engines at 100 percent for the "cat shot." Wham! The catapult fired. Off I went into the night and quickly entered the carrier pattern downwind leg. I flew two perfect passes and made smooth landings, probably because I could not see well enough to screw them up. The weather was beautiful and the carrier landings were easier than the field carrier practice.

World Cruise (September 1953 – April 1954)

Our next scheduled deployment was a long "world cruise" for September of 1953. For nine months prior to our departure we engaged in ordnance training, air-to-air gunnery, rocket firing, and bomb delivery. The training period between the long cruises was not entirely spent flying from the home base at Jacksonville. We had numerous short cruises and deployments, almost to the point that we were looking forward to a long cruise. Sometimes we were deployed with scant notice. Usually we had a little warning, but it was not uncommon to come home and pack a bag and head for Cuba for two weeks. It was discouraging going to work in the morning without a guarantee that you were coming home that night.

Ordnance training was interesting because we were carrying maximum combat weapon loads and dropping them on trucks and other vehicles on the target range at Pinecastle, Florida. One of our ordnance configurations included two 500-pound bombs. When they exploded, the shock wave was visually impressive. We also carried a mixed load with bombs and rockets. A pilot had very careful with his altitude while releasing various weapons. One of division leaders, "Black Mac" McClosky, inadvertently released a 500-pound bomb at low altitude. He had planned to fire some 5-inch High Velocity Aircraft Rockets (HVAR), but apparently his selector switch was in the wrong position. Mac's plane got a belly full of his own bomb shrapnel and he almost didn't get his Banshee back home.

During the summer of 1953, one of the more senior squadron pilots, Leroy Ludi, completed his four-year tour in the service. Audrey and I had become very good friends with Leroy and his wife during our time together in the squadron. Leroy was a smart fellow. He headed for the University of Florida to finish his college degree in engineering. Leroy and I decided to keep in touch and eventually we ended up working together. September drew near and so did Audrey's due date for the baby. We decided that after the baby was born, Audrey would go back to Pennsylvania to stay with her parents. In the meantime, I was headed back to the North Atlantic for a long cruise and Audrey's parents came to Jacksonville.

I departed Norfolk on the U.S.S. *Wasp* (CV-18) in early September. I felt a little better on this second trip, because of my earlier experience, but it was still very difficult flying. We had a little extra surprise waiting for us on this trip: fog. According to Navy regulations, we did not launch jet aircraft if the temperature and dewpoint of the air were within three degrees of each other. Our jets were limited on fuel and endurance in the air. With a dewpoint that close, fog formed very quickly, making it impossible to land on the deck. A jet aircraft might run out of fuel waiting for deck visibility to clear. Propeller aircraft, primarily the AD dive bombers and sub hunters had a lot more endurance and could be launched with a close dewpoint, as long as there was no fog at the time of takeoff.

During this cruise, the U.S.S. *Wasp* participated in a North Atlantic Treaty Organization (NATO) cold weather training operation named Exercise MARINER. Other participating aircraft carriers included the U.S.S. *Bennington* (CV-20), the Canadian H.M.C.S. *Magnificent*, and the British H.M.S. *Eagle*. They operated as the core of the fleet, surrounded by numerous destroyers and cruisers. The task force included over 300 ships and 100 aircraft.

On 23 September, I was in the number two Banshee on the catapults waiting for launch with engines running. I noted that the deck crew directed the Number One Banshee to taxi onto the forward elevator and cut his engines. He immediately dropped out of sight as the elevator lowered him down to the hangar deck. I assumed that he had a fuel leak or some problem, but that was not the case. They had me taxi to the elevator next and down I went. The other jets soon followed. The temperature/dewpoint separation had dropped to two degrees and all jet flights were cancelled.

Meanwhile, the carriers launched their propeller-driven aircraft. Less than one hour after the launch, the dewpoint spread went to zero and we were socked in. We had zero ceiling and zero visibility. The carriers notified the airborne crews of the situation below and told them to come back to the vicinity of the carriers and to set a maximum-endurance power setting on their engines. Now we had 42 aircraft and 63 airmen circling above in the sunshine while their carriers cruised unseen in the thick fog below. The estimated endurance of the aircraft was set at five hours and everyone believed that surely the fog would lift some by then. After an hour or so of waiting, the operations officer decided to allow the radar-equipped ADs (called "Guppies") to attempt instrument approaches to the carriers on their aircraft radar. There were no precision instrument approach systems available on carriers at the time.

After my flight was canceled, I returned to the ready room and followed the operation from there. I went up to Pri-Fly one time, but there was nothing to be seen up there. The fog was so dense, I could scarcely see my hand in front of my face. Since there was nothing to see, I went back down to the ready room. When I heard the ADs were going to attempt approaches, I returned to Pri-Fly. I did not believe that any aircraft could find its way to a landing in this "pea soup." I turned out to be right. The AD pilots had been briefed not to descend below a minimum altitude for fear of flying into the water or hitting the superstructure of the carriers. They had radar altimeters to give them an accurate altitude over the water. Initially, it was relatively quiet on the catwalk outside of Pri-Fly. Soon, I heard a slight drone of an AD's engine out in the fog. It grew louder and louder and all at once it seemed to be right on top of us. We never saw the AD but it passed almost directly over the ship at low altitude. All of the observers and "lookie-loos" immediately cleared the Pri-Fly area and returned to the ready rooms. This was a dangerous operation and an approaching airplane could easily strike the carrier.

After several scary approaches the AD pilots reported that they never saw the carrier. We never saw them either. About three hours into the ordeal, the Operations Officer called off the approaches and put the aircraft back into the maximum-endurance holding patterns above. Tension built-up and a sense of fear and foreboding coursed through the ship's company. The carrier put out a call to any ships in the area for weather reports. We needed at least a 75 to 100-foot ceiling to allow us to recover our aircraft. The submarine U.S.S. *Redfin* (SSR-272) heard the call and surfaced about 100 miles west of the carriers. He reported that his location had about a 1,000-foot ceiling over the ocean with two miles visibility and rough seas. The carriers immediately took a course toward the submarines position, but that was nearly four or five hours away and the aircraft were going to run out of gas in less than two hours.

Operations decided to send the aircraft to the vicinity of the submarine while they still had enough fuel to find the sub and ditch beside it. That way perhaps half of the crews would be recovered from the stormy surface of the North Atlantic. That was our best-case estimate. Half the men would probably die. The mood aboard ship was one of complete despair. The Captain briefed

The Smell of Kerosene

the ship's personnel on the situation, using the public address system. Next, the chaplain came on the air and said a prayer for the flight crews. Everyone on the ship prayed with him, though there seemed little hope of a happy outcome. I have to confess that I had given up any hope myself, but I was thankful for my luck. If I hadn't been a jet pilot, I would have been up there with them.

About 20 minutes after the aircraft headed toward the submarine, the carriers entered an area where the fog was a little higher with a nearly 300-foot ceiling. The aircraft were immediately recalled to the ship and the "Guppies" were directed to fly alongside the non-radar-equipped airplanes to guide their approaches. I went back up to Pri-Fly with a number of other pilots to observe. We could tell that the situation had improved, but landing conditions were still extremely poor. All at once, two ADs appeared out of the mist behind the carrier's wake. One had his gear and flaps down for landing. The other, the radar plane, guided him in. As soon as the landing AD pilot called the carrier in sight, the Guppy pulled up into the fog and climbed to pick up another aircraft to bring home.

The LSOs were very generous and gave no wave-offs. If the AD was near the fantail of the carrier, he got a cut and came aboard. There were no bad landings or incidents. Recovery operations continued for nearly 45 minutes on all of the carriers. A Canadian Sea Fury fighter landed on the *Wasp* and some of our ADs landed on the *Magnificent*. No one cared where they landed as long as they got aboard some carrier. When the last Guppy landed and all aircraft were back onboard, a cheer went up over the ship. The chaplain came back on the P.A. and thanked God for answering our prayers. I don't think there was one non-believer on any ship in the fleet that day.

The flight surgeons broke out the medicinal alcohol (brandy) to celebrate and help relieve the stress. The fog lifted sometime the next day but it was several days before the "whiskey front" cleared. The pilots and aircraft that landed on carriers other than their own waited until the next scheduled air operation, about a week later, before they returned to their home carrier.

In spite of some rough seas, stressful flying, and accidents, my second North Atlantic cruise provided one of my happiest moments. On 29 September 1953, I received a telegram notifying me that I was the proud father of a baby daughter named Sandra.

Close Call

I experienced one close call that I knew nothing about until after the mission was over. All jet takeoffs from the carriers were by catapult launch. This was always an exciting ride that went from a static condition to about 140mph in a distance slightly over 125 feet. When the catapult fired, the high acceleration pressed my head and body back to the seat forcibly. At this point, I was just along for the ride. The aircraft was pre-trimmed and the stick held in the neutral position allowing the aircraft to fly on its own. The controls also had a hand or finger hold, allowing me to prevent the throttles from creeping back during the acceleration. As the aircraft departed the flight deck, acceleration forces suddenly decreased. I retracted the gear and configured the aircraft for climbing flight.

I was waiting for my "cat shot" one day when the crew chief forgot to secure the main fuel cap on the fuselage, just behind the canopy. When the catapult started its acceleration stroke, the fuel was forced to the rear and top of the tank. Because of the missing cap, a large cloud of jet fuel vented back over the aircraft and the hot engine exhaust. After we had recovered from the mission some hours later, one senior pilot told me he had only witnessed one other similar situation. In that case, the aircraft exploded in a ball of flame right in front of the ship. In my case it didn't and God only knows why not. I guess He was riding with me that day and on many to follow in my career.

The mission was one of our many "show of force" demonstrations. During an airshow over Beirut, Lebanon for the local U.S. Ambassador and other VIPs, we conducted high-speed low-altitude passes, vertical pull-ups, and inverted flight. Somehow, the missing fuel cap didn't cause any noticeable problem for my aircraft. None of the other pilots in the flight noted any fuel escaping from my aircraft. I was a little upset that the operations people in Pri-Fly had not called me on the radio to report the spray cloud or loss of fuel during the "cat shot." After I returned, a lot of people welcomed me back and described the big cloud of fuel behind my aircraft that almost blanked it from view, but not one report was made at the time. This one just slipped by. It should have been reported. I determined that I lost 500 pounds of fuel during the flight. The chief of maintenance later informed me that the crew chief had been disciplined.

Combat Practice

Carrier flying wasn't all stress punctuated by moments of terror. It was often exciting, fun, and a source of great pride. I derived a tremendous sense of satisfaction from performing to the best of my abilities and contributing to an efficient operation. I was proud to be part of an elite team. Most often, flying provided a view of the heavens and the ocean that few other than naval aviators ever experienced. I can vividly recall the thrill of spotting the fleet after returning from a mission hundreds of miles away. The small white wakes of the ships were a welcome sight after flying over miles and miles of empty ocean. I always savored the final "cut" and landing as I came aboard with another mission and carrier landing under my belt.

During the voyage, we conducted numerous simulated combat operations. From the North Atlantic, we launched simulated airstrikes into Germany and other areas just as if we were at war. We flew against aircraft from Britain and Norway that represented our potential adversaries, Soviet interceptors. Our air-to-air duels with these other aircraft were thrilling. We did not have instrumented tracking ranges like those at Nellis AFB, Nevada that track and record data to confirm success or failure in the engagement. We had gun-camera film to verify our tracking, but no one could tell if we were taking good pictures of our supposed adversary sometime after his buddy had already shot us down. It was an exciting time, twisting, turning, climbing, and diving, always trying to get into the firing "saddle" position on an enemy aircraft.

Our opponents flew de Havilland Vampire and Gloster Meteor jets from England. The Vampires were pretty little single-engine fighters with twin tail-booms. They didn't have the performance of the Banshee. The early Meteor F Mk.3, the ones we faced the most, were about an even match for the Banshees and made the air-to-air maneuvers an interesting challenge. There were a few late-model Meteor F Mk.8 aircraft that had bigger engines and were easily identified in flight by the large nacelles on the wing. We never toyed with these unless we had a good starting advantage because they would eat up the Banshee. I can recall joining up on the wing of one of the Meteors and noting that the pilot was wearing a cloth WWII-type flying helmet. I was surprised that they didn't have the hard-shell crash helmets that we were issued in the Navy.

After sorties engaging the "enemy" fighters near the coast of Europe, we scheduled a deep penetration into Germany to make simulated strikes on a railroad yard and fuel tank farm. The plan called for the Banshee because of its tip tanks and extra range. We flew at high altitude until we approached the point where the enemy fighters were patrolling. As we approached to within about 20 miles, we went into a high-speed descent to the deck and flew across the coastline unmolested on our way to the target. I'll never forget how green and beautiful the farms were in northern Germany. The red barns stood out along with the white farmhouses. Lakes and pastures made a picture-postcard background. When we located our target, we made a series of high-speed, low-altitude runs over the rail yards and oil tanks just to let the natives know that we were in town.

Back to the Med

We completed flight operations in the Atlantic and entered the Mediterranean Sea. The weather was much nicer, and we settled down to cross the Med from the western entry at Gibraltar to the eastern departure through the Suez Canal. Our primary destination for this world cruise included a tour of flying in Korea. The cruise had been scheduled in January 1953 and our preparation from that date until our deployment had been directed toward flying combat in the Korean War. In July 1953, however, a truce was signed between North and South Korea and the need for our squadron to fly in combat was eliminated. The cruise was not cancelled, however. Our new mission was peacetime patrol in the Korean area. I recall a number of my squadron mates vocally sounding off at how disappointed they were that we were not going to get into combat. I had mixed emotions about this. I took pride in being proficient at delivering bombs and rockets and I realized that this was the primary reason the Navy had trained me, but I could not honestly say I was disappointed by the fact that I would not have to drop my weapons on people. The role of my squadron in that sort of a war was tactical air-to-ground support. The Air Force had the responsibility of handling air-to-air combat against the MiG-15 fighters used by Soviet, North Korean, and Chinese pilots. The Air Force pilots accomplished this task with the F-86, a fine airplane.

We spent little time operating in the Mediterranean on this cruise and we entered the Suez Canal in November 1953. At the time, the carrier *Wasp* was reported to be the largest ship to sail through the canal under its own power. There was not much clearance between the sides of the ship and the canal on most of the passage through. I remember the warning given to all ship's crewmembers not to stand in front of lighted passageways during the night on our trip through the canal. There was some risk of bored Arabs in the surrounding desert taking pot shots at us with their long rifles just to break the monotony. One evening as we were slowly making our way through the canal, we felt a solid bump or jolt on the ship and then nothing more. The carrier continued her passage. We learned in the morning that a smaller ship with several Arabs aboard was illegally using the canal in the opposite direction at night and had collided with the carrier and sank right behind us. We were happy to clear the canal and get out to sea and back into air operations.

We arrived in the Sea of Japan in December 1953 and joined another carrier operating off the coast of Korea. The *Wasp* stayed on patrol for a month to six weeks while another U.S. carrier went into port in Japan. After a little shore leave and replenishment, the other carrier relieved us and we got some time in port. Our ports included Yokuska near Tokyo and Itazuki in southern Japan and we were usually in port for about two weeks. Occasionally, there were times when both carriers were at sea and operating together.

During one of our port visits, we had one of those large "screw ups" that occur at times in big operations. It was common practice to offload jet engines and other major aircraft components that were out of service and on their way to an overhaul facility by other Navy transport ships. At the same time, the new or replacement items were loaded on the carrier for its next patrol at sea. The jet engines were stored in olive-drab steel containers that were referred to as "engine cans" and it was normal to see four or five of these sitting on the dock just abeam the carrier. Not many people were aware of it at the time, but apparently the special nuclear weapons that I mentioned earlier were also stored in cans similar to the engine cans. One of these had been erroneously offloaded on to the dock along with some jet engines. No one was aware of the error, and so the weapon sat on the dock along with the out-of-service jet engines.

We returned to sea on our next scheduled patrol and, after about one week of normal operations, a nuclear practice drill was called and the ship went into its high-security mode. As I heard it, once the special weapons locker was opened, it was quickly discovered that one of the "special weapons" was missing. There was much consternation among the personnel involved in the breach. Fortunately, the container was found sitting on the dock with the out-of-service engines waiting for the transport ship that had not yet arrived. The weapon was not a safety threat because all sorts of special arming devices, required to detonate it, were normally not set until completion of final loading on an aircraft prior to launch. Nevertheless, it was very embarrassing for the Navy.

Southeast Asia

In March 1954, U.S.S. *Wasp* was back on patrol again. The task force cruised in a racetrack pattern near the northern end of the Sea of Japan. The weather was cold with rain and snow, and I went up to the Pri-Fly and bridge just to observe the radar. No other ships were in sight, but the bright blips on the radar screen indicated that they were in position. A carrier normally did not operate without a supporting force of destroyers, and sometimes a cruiser or two. It was night and there were no flight operations. I was glad that the night fighter pilots didn't have to fly on a miserable night like this. After talking to the officer of the deck and watching the radar scope, I went back down to the wardroom for coffee and chatted with several other pilots who were at "loose ends." An hour or so passed, and the vibration in the carrier increased, hinting at an increase in speed similar to that which occurs during flight operations. Another pilot came to the wardroom from the bridge or Pri-Fly and commented that the carrier had departed the racetrack pattern and was now on a heading of south. Initially, not too much was made of this because the Navy had a way of doing things to prevent outside observers from predicting its actions. However, after several hours on the southerly heading at the higher speed, we knew that something was in the air and we were heading somewhere new.

What had occurred was that we were on orders to depart the northern Sea of Japan and proceed to the South China Sea off the coast of French Indo-China, or Vietnam as it is known today. The French had been waging a war with Vietnamese insurgents for some time in this area, and the U.S. had been providing warplanes and supplies to the French. The North Vietnamese had put a final push on the French and were threatening to drive them out of the area. The U.S. was ferrying supplies to support the French Air Force. At this time, several squadrons of Chinese MiG-15s had been moved to southern China where they were in position to intercept the transports in their re-supply effort. There were no fighter bases within range to support the transports, and the U.S. Navy responded by moving the carrier task force into position near Vietnam.

We started immediate intelligence briefings on the MiG-15's performance and tactics. What had only been a distant possibility of encountering MiGs when we were patrolling in Korea was now a distinct possibility. Every step of our plans and preparations was directed toward this end. The rocket and bomb rails that were attached to our wings for the fighter-bomber role were quickly removed and the holes and gaps taped over with the heavy green aircraft tape. Any item of excess weight that was related to the ground-attack role was removed. This gave us as much performance as possible for the air-to-air mission. The MiG-15 was a swept wing aircraft, and had an edge on the Banshee for speed, but there were areas of altitude and speed where the Banshee could outmaneuver the MiG-15 in turning performance. We reviewed all the advantage areas that were ours, and we knew what arena we had to get the MiG into to beat him.

We were ready to go. I had never been more excited in my preparation for an assignment. This was one that I was looking forward to. It was a joy to see the Banshee stripped of all of its air-to-ground weapons system, and set for the fighter role. We stayed on patrol for three weeks waiting for the MiGs to venture south over the Chinese border toward the stream of transport aircraft bringing supplies to the French, but they never came. We had combat air patrols flying almost continually, and when we didn't, we had manned aircraft on the catapults. They were preflighted, armed and ready to launch within two minutes. This period was one of high excitement and expectations, and I personally was confident that we could engage the MiGs successfully. Our squadron had experienced good success in simulated air-to air engagements with the Navy's swept wing F9F-6. The MiGs were lighter and had better performance than the F9F-6, but that did not discourage us a bit. Rear Admiral Whitney, Commander of Task Force 77 used *Wasp* as his flagship. He would periodically appear on the flight deck to talk to the pilots on catapult alert. I can recall his presence and his attitude. He didn't flaunt his rank. There was no pomp-and-circumstance. It was one man to another. If the MiGs came south, we went after them. The orders were in place.

At the end of the three-week period of watching the MiGs circle above the border without venturing toward our air transports, our pilots were getting a little anxious and making noises about going after them where they were. About this time, the French and their supporters decided to abandon French Indo-China and they pulled out. Our task force headed back north to the area around Japan, and I was one of the pilots who vocally offered regrets that the chance for engagement had gone by. I'm sure that our presence in the region was a deterrent to the Chinese and made the transport pilots feel better.

The task force returned to Japanese waters and, shortly thereafter, departed for San Diego, California. Our nine-month world cruise was coming to an end. Although it was an experience I would never forget, I was ready to get home. I wanted to see my wife and, for the first time, my nine-month-old daughter.

Heading Home, April 1954

The crossing from Japan to San Diego, took about ten days, and the carrier served as a transport for not only our aircraft, but others that had been damaged in flight operations and were being returned for overhaul and repair. Because we were headed for home and not flying, the journey passed slowly. Midnight movies in the ready rooms and poker games in various staterooms helped pass the time.

Once back in the States, I returned to Jacksonville. I took leave and went to Pittsburgh to pick up my wife and daughter. Because Sandra had been born only weeks after I left on the cruise, this was my first time to hold her and really experience the feelings of being a father. I spent an enjoyable visit with all the family and friends, but it was soon time for Audrey and me to pack up and get back down to Jacksonville to find an apartment. I still had a little duty time until September 1954.

I definitely enjoyed Navy carrier flying. It was demanding, challenging and satisfying. It gave me a great deal of pride and confidence that I was able to do it well and be a part of such an important operation, but weighing all the factors involved, I decided to sign out after my four-year tour was over. The extended sea duty and separation from my family was just too difficult and my constantly changing schedule was a problem.

From May to September 1954, I became more and more of a "short-timer." After September, I would be gone from the squadron and unavailable for cruises. The newer pilots were getting the priority on flying and training, just as I had when I first entered the squadron. The cycle was repeating. I had been in contact with my old squadron mate, Leroy Ludi, who had left about a year earlier and enrolled in the University of Florida at Gainesville. He was also flying in the Navy Reserve at Jacksonville and this helped toward his schooling.

Service in the Navy Reserve appeared to be a very viable option to extending my service in the Navy. I put in my application for the university during the summer of 1954 and was accepted to enter the fall term starting in September. I moved from Jacksonville to Gainesville, and started class, all on the same day.

Florida on the G.I. Bill

Because of my experience in Naval aviation and my desire to pursue a career as a test pilot, I enrolled in the Aeronautical Engineering department. Several of my friends left the Navy about the same time and found jobs flying with the airlines. I never considered this option. Test flying looked like the most interesting challenge for me. In addition, I wanted to finish college and get a degree. My studies went much better than my earlier years at Penn State. I had only been an average student then. By the time I re-entered school at the University of Florida, I had matured and gained a great deal of experience and self discipline. Most importantly, I was able to read and study in a much more efficient manner. The Navy speed-reading lessons during pre-flight training really paid off. In addition, I was married, with responsibility for a family, and schooling was important. It was a stepping stone into my desired field.

Financially, we were doing well for the average college family. I was on the G.I. Bill that was established for veterans and I was flying in the Navy Reserve at Jacksonville. The reserve flying had a double benefit to me. It provided extra money for living and school but, more importantly, allowed me to keep current in flying jet aircraft while I was finishing my degree. The minimum qualification requirements for a test pilot included military flight experience and an engineering degree. In fact, my Aeronautical Engineering degree would later prove extremely valuable.

During my second year at the university, Audrey had become bored with so little to do in our small apartment. She found a job with the Florida Welfare Department. About the same time, I took a part-time student engineering assistant job in the Mechanical Engineering department at the university. The pay was not much, but the experience was good and the work related to several U.S. Air Force contracts with Eglin Air Force Base on an F-86D flight test program. This made it more interesting for me.

Naval Reserve

Reserve flying at Jacksonville had another benefit for me. It was my chance to get away from the rigors of study and school for one weekend a month. Jumping into the F9F-6 Cougar and leaping into the air was certainly a good way to clear the cobwebs of extended study from my mind. It "recharged my battery" and, after a weekend of flying, I was in much better shape to go back and hit the books. In the summer, the reserve called us up for two weeks of active duty for a little more extensive training. A little adjustment to my summer school schedule allowed this to work in well, and the two weeks provided a little nest egg of cash for the fall semester.

I had applied for a billet in the Navy reserve squadron at Jacksonville prior to leaving active duty, but I had to wait for an opening. The nearness to Gainesville and the participation of ex-Navy personnel now in college kept the reserve squadrons at Jacksonville full. However, as students graduated and accepted jobs around the country, there were openings for people who were just entering college. I joined reserve jet fighter squadron VF-741. The squadron had an interesting cross-section of pilots. About one-third were college students and others were established in their civilian careers. We had lawyers, engineers, contractors, salesman, technicians from the various trades, and even a few farmers. The squadron had been flying a World War II vintage fighter, the F4U Corsair, for some years. A number of our pilots had fought in that war as well.

I truly enjoyed the camaraderie and friendships that came along with being a member of the reserves. This was a diversified group of people who, in essence, belonged to a unique private club who met monthly to fly and communicate with each other. The older F4U aircraft were replaced soon after my arrival with the F9F-6 Cougars. I'm glad I had the opportunity to fly the F4Us for several flights. I had not flown a big reciprocating engine propeller aircraft since the spring of 1952. All of my current time was in jet aircraft. The first thing that I forgot was how to start the big radial engines that have personalities of their own.

There was also a Marine reserve squadron at Jacksonville and they flew and maintained the Corsairs. We drilled on different weekends, but flew the same airplanes. I read the handbook, completed the written quiz, talked with the older Corsair pilots and prepared to fly. After I strapped in, I received the all-clear signal from my crew chief, a Marine sergeant standing by the fire bottle (extinguisher). I hit the starter switch and the big old four-blade prop slowly turned. After about eight revolutions, I turned the magneto switch on and began to prime the engine. That old engine burped, snorted, and fired, but it would not start. After some time, I noted the look of anguish on my crew chief's face as he watched his airplane and worried about his starter overheating or burning out. That was enough for me. I stopped cranking the engine and motioned for the crew chief to come up on the wing. He did so, appearing at the cockpit almost immediately. I looked at him and said, "Sergeant, if you will start this old girl for me, I will fly her for you." He put on a mile wide grin and reached into the cockpit. He hit the starter and primer, moved the throttle to just the right spot, and had that old R-2800 purring. Even though I had forgotten the technique, I knew there was one, and I knew the crew chief would know it well.

Because I was only going to have a few flights in the Corsair, I put it through its paces. I wanted to see what the pilots of World War II had to work with. It was a great airplane to fly and as I passed between the cumulus clouds around Jacksonville, I mentally pushed the time back 12 years and waited for a Zero pilot to show his face. The jets had spoiled me somewhat in that the control forces seemed high in the Corsair. I'm sure that, during the war, pilots had adequate adrenaline to overcome this when they engaged the enemy. The airplane responded well and that is what counted. After a little inverted flight that is always part of the fighter pilot's world, I noted the smell of gasoline in the cockpit that indicated the old carburetors didn't like to stay upside down too long. As I headed back to the pattern for landing, I had to marvel at the old "Hose Nose" as she was frequently called.

Flying the Cougar

The replacement aircraft, F9F-6 Cougars, soon arrived. These airplanes were newer and had more performance than the F2H-2 Banshee that I had flown in the fleet. The transition to this aircraft was easier for me than the short transition that I had made to the F4U. For some of the older pilots in the squadron who had flown Corsairs since WW-II, the transition to jets was a major event. Some decided not to make the transition at all, and instead transferred to another reserve squadron that had the Grumman AF-2 Guardian anti-submarine aircraft.

I was one of the squadron instructors, a pilot with a lot of jet experience who trained pilots who just had a propeller background. One problem that we had with the "old heads" was getting them to fly the jets at the high speeds that they were designed for. We often found the older pilots lapsing into F4U climb speeds of 250 knots, and we had to remind them that they were now operating in the jet world and the climb speed was 400 knots. Considering that this was a large change for some of the older pilots, the transition to the Cougar went well.

The Cougar was a typical Grumman product. It was one of the strongest fighters ever built. A pilot could make practically any control input and maneuver at nearly any speed without concern for structural failure. This was a new experience for me, as the old Banshee was a little sensitive to loading and the wings could tear off in a high "g" situation. The Cougar also had nice stability characteristics. If it "departed" (went out of control) at high speed, the pilot merely had to release the stick and it recovered itself quickly.

There was one fun maneuver in the Cougar that I always enjoyed as it was a real confidence-builder. I would accelerate the Cougar to Mach 0.9 at 30,000 feet and pull the stick back sharply while applying full aileron or spoiler. The aircraft then snap rolled and departed controlled flight. By just releasing the stick, the aircraft recovered quickly, usually inverted, at a little less than 1g. I usually found myself hanging upside down in my harness with the sky seemingly "below" me and the ground "above." It was a spectacular maneuver and one that gave me confidence in my aircraft.

When I flew my reserve training on the weekends, I had an assigned syllabus of training flights to complete. This included instrument training, formation and tactics. I also had time to challenge or "jump" targets of opportunity, including fleet pilots who happened to be flying on weekends or Air National Guard aircraft out of Jacksonville Municipal Airport. This latter type of flying simulated the real world more than the "canned" training. I quickly gained a level of respect for the Air Guard who, at the time, were flying the World War II vintage P-51 Mustangs. The Guard had just returned from a tour in Korea, and the pilots were experienced. Even with my higher-performance jet, it was hard to catch a P-51 pilot unaware. About the time we were entering the pursuit curve or were within simulated firing range, the P-51 would pull a tight turn into us and we would flash by. As you might imagine, I was at Reserve duty more than studying, but both were leading to a career in test flying.

Looking for Work

When I entered my senior year in the fall of 1956, I wrote letters seeking employment. I wrote to the chief pilot of every large flight test organization that I could find. I didn't care where they were located. The job was my primary consideration. At the time, new graduate engineers were in short supply and in high demand. Almost all the major aircraft companies were interviewing at the campus. I had several interviews and received good offers to join the companies in their flight test department as an engineer. The offers hinged on two things: grade point average and background experience. The majority of the responses that I received were cordial, but negative. There were no test flying slots available. A number of company chief pilots indicated that they selected their pilots from qualified pilot-engineers in their own flight test departments when the need arose. I received one really bad, almost insulting, response.

I finally received a positive response from Jack Reeder, Chief Pilot at the National Advisory Committee for Aeronautics (NACA) Langley Memorial Aeronautical Laboratory (LMAL) in Virginia. Reeder wrote that they would be looking for a new pilot around June 1957. He also indicated that he would like to interview me. I bought an airline ticket and headed to Langley. My old Navy squadron mate, Leroy Ludi, was already working at LMAL as a helicopter research engineer. Apparently when I interviewed with Jack Reeder, I didn't seem as excited about the prospect of flying for him as he might have expected. I suspect that was due to my attempt to play it cool. I really was excited as I observed all of the interesting projects they were flying. In any case, Jack indicated that they would notify me in two weeks, one way or the other. He also told me there were other candidates being considered. Later, Leroy told me that Jack asked him if he thought that I really wanted the job. Leroy set him straight, telling him that I was very interested in the job and really wanted to fly at Langley. I know that Leroy's input was very important.

The two weeks seemed to pass very slowly, but in about ten days I received the offer by telegram. I was thrilled. I knew from all of the responses that I had received that there were few openings around the country for test pilots. I felt extremely fortunate being selected by the NACA. I was headed directly from college into a test-flying slot without first working as a flight test engineer. After I had received and accepted the offer, I mentioned it to one of my Aeronautical Engineering Professors, Bill Miller. When Miller, a retired Navy captain, heard where I was going, he exclaimed, "Don, I wish you had told me sooner. I am a personal friend of Mel Gough." Gough, another retired Navy man was Jack Reeder's boss, and the head of the NACA Flight Operations Division. Sometime after I had checked in, Jack told me that Professor Miller had written a very positive follow up letter to Mel, congratulating him on his selection. Jack was pleased in that it reinforced his decision. The next closest candidate under consideration was another ex-Navy pilot who was working for Proctor and Gamble. His degree was in Chemical Engineering, and I'm sure my Aeronautical Engineering Degree was a deciding factor.

3

NACA and NASA – Langley, Virginia

Research and Support Flying

The National Advisory Committee for Aeronautics (NACA) was created in March 1915, as a rider attached to the Naval Appropriation Bill, to develop and nurture the budding field of aviation. Overseas, the importance of aeronautics had already been recognized. Dirigibles and airplanes quickly earned a prominent place in the war in Europe. The NACA sought to close the gap with a mandate to "supervise and direct the scientific study of the problems of flight, with a view to their practical solutions." Within two years, the NACA established its first independent laboratory. Named in honor of Samuel Pierpont Langley, former Secretary of the Smithsonian Institution, the facility occupied part of a U.S. Army airfield (later called Langley Air Force Base) across the river from Norfolk, Virginia.

When I reported to Langley Memorial Aeronautical Laboratory (LMAL), there was no way of knowing that within a year the Soviets would launch the first earth satellite. On 4 October 1957, Sputnik would cause a response in the United States that changed the nature of the NACA forever. The Committee, along with its facilities and scientists, would form the nucleus of the National Aeronautics and Space Administration (NASA). My career spanned one year with the NACA and 29 years in NASA.

When I accepted the job at Langley, I entered as a GS-11, entry level for test pilots. The equivalent level for a starting engineer was a GS-7. A GS-11 earned a salary of $7,700 per year. The offer that I had turned down for a flight test engineer's position at McDonnell Aircraft was $8,400 per year. There was a significant difference in government and industry salaries, and it remained that way as long as I worked for NACA and NASA. Money was not my primary consideration. I just wanted to fly. I was extremely happy to get the job for $7,700 per year and I probably would have accepted $6,600 per year if that had been offered.

The Langley flight operations section was staffed with five pilots when I arrived. I was the sixth. Jack Reeder, the chief pilot, was one of the few at the lab who did not have military flight experience. He had served as a wind-tunnel engineer at LMAL during World War II. He also had a Civil Aeronautics Authority (CAA) pilot's license and had a reasonable amount of private aviation experience. During WW II, NACA Langley introduced a small in-house program for training civilian aviators as test pilots, since military pilots were

NASA photo L-78005

The NACA Langley Memorial Aeronautical Laboratory (later NASA Langley Research Center) was the agency's first independent laboratory.

involved in the war effort. Jack had done extremely well in the program and had progressed to the position of chief pilot. He had also checked out in high-performance jet aircraft and flew test programs on them. Jack's real expertise was in vertical takeoff and landing (VTOL) and very short takeoff and landing (VSTOL) aircraft. He had the necessary abilities to operate and test some of the most unusual and difficult-to-fly VTOL aircraft that existed in the early years of their development.

Jack was a tough taskmaster to work for. One of his first engineering assignments to me was to write a NACA Technical Note (engineering report) on helicopter evaluation program. The project had been passed around between two or three of the other pilots, but had never been completed. After a lot of research involving scarce written records and interviewing the other pilots involved, I decided to have the helicopter returned to the test configuration and conduct some tests of my own to provide additional data for the report. I then presented my first rough draft of the report to Jack for his review and critique.

Such review and scrutiny was a normal part of NACA operations. When a report was printed by the NACA, you could be confident of its accuracy and the validity of the data. The NACA earned this reputation over the years and it was respected throughout the aviation industry. When Jack returned the rough draft to me, he made a number of perceptive comments and suggestions for improvement. I answered these to the best of my ability and re-submitted the report to him. The report came back with general acceptance of the area that I had corrected, but I found that Jack had changed areas that he had accepted on the first review. This went on through about five reviews and, toward the end, I became quite exasperated.

Jim Whitten helped me a great deal with my report and told me that this was typical of Jack and his reviews. I was pleased to discover that after the five or six different rewrites, the final rewrite that Jack accepted was closer to the original than any other. We had come full circle, so to speak. I submitted the report to the editorial committee and there was no additional comment. I could not resist showing Jack the first draft as it compared to the final report and he only sort of grumbled under his breath. I think he was purposely training me in the nuances of the NACA reporting system and the importance of accuracy. To me, it just proved that there were several ways to tell the same story. It was a matter of personal preference.

Jim Whitten was a former U.S. Air Force pilot and a colonel in the Air Force Reserve. Jim was a true southern gentleman. He was over six feet tall and carried himself well. His appearance and demeanor earned him a quiet respect. Jim, the next senior pilot in the office after Jack Reeder, was in the twilight of his flying career. During the six years that I worked at Langley, I watched his eyesight slowly deteriorate until the flight surgeons took him off flying status. He was a fine aviator and I consider it a privilege to have had him as a fellow pilot and friend.

The other three pilots in the office were former Naval aviators. They were Bill Alford, Bob Champine and Robert Sommer. Bill and Bob had some seniority in the office. Robert, however, started two years before me and was the "boot ensign" until I arrived on the scene. As ex-Navy pilots, having gone through similar training and experience, we related to each other quite well. As the fledgling test pilot, the others in the office were like instructors to me. Even though my experience was growing with time and I had the engineering degree and military flight experience, this was my introduction to flight testing and I accepted every bit of help that I was offered.

Bill Alford was the senior ex-Navy type. He had originally been a Navy PBY patrol plane pilot. He flew everything at Langley and evaluated aircraft at other test installations. Bill was my mentor, although there was no official letter or document assigning him the task. I suspect that Jack Reeder just walked through the office one day and said, "Bill, look out for Mallick and see that he gets checked out." That was the way it was done in those days. There was little or no formality, but it worked well. I couldn't have had a better instructor. Bill and I got along great and eventually shared some flying projects together.

The Smell of Kerosene

NASA photo L-46931

The Grumman JRF-5 Goose served as a support aircraft at Langley. The versatile amphibian was mainly used to ferry cargo and people between Langley and the Wallops Island rocket test station.

July, August, and September of 1957 were probably the busiest flying months that I ever experienced in my career because Langley had a lot of research projects going on. I was checked out in various aircraft to help with support flying, such as safety chase for research missions. This allowed the other pilots to concentrate on their test programs. I checked out in three multi-engine aircraft: Lockheed Model 12A Electra (similar to Amelia Earhart's), Douglas C-47B "Gooney Bird" cargo plane, and Grumman JRF-5 Goose amphibian. These aircraft were used to carry passengers and cargo for Langley; NACA Headquarters in Washington, D.C.; and Wallops Station, which conducted rocket testing over the Atlantic Ocean near Chincoteague, Virginia. During this time period, I was also checked out in helicopters because Langley was involved in several rotary-wing test programs. Considering that I was checked out as airplane commander in several multi-engine aircraft, learning to fly helicopters, piloting an amphibian, and flying several jet aircraft, I was extremely busy that summer. Learning to fly so many new types of aircraft in a short period of time was a great experience and a terrific confidence builder.

Helicopters were a completely new experience for me. It was different from any flying that I had previously done. Early in my Navy career, I had unconsciously looked down at helicopter pilots. I thought that anyone who couldn't fly in jet aircraft must be among the downtrodden. It didn't take long for me to learn the truth and suffer a little humiliation when I tried to fly a helicopter for the first time. As a hotshot ex-Navy jet pilot, I discovered that rotary-wing aircraft were a real challenge. Helicopter pilots required special capabilities and skills that sometimes went beyond those required for jet flying. The coordination and techniques required to fly helicopters are unique, and not all pilots have them. I discovered that I did, but I had to work for it and apply all of my skills in order to do an adequate job.

I flew with all of the pilots in the office during my checkout in various Langley aircraft and helicopters, and it was beneficial for me because all were highly qualified and expert in certain areas. Jack Reeder normally signed me off for helicopter qualification. Occasionally, certain other pilots did it in his absence. I began to truly enjoy flying helicopters, but it always required an extra degree of concentration. I couldn't let up or relax too much. Most helicopter pilots approach flying in that manner if they intend to survive.

When it came to the Grumman JRF "Goose," a small Navy amphibian, Bill Alford performed the checkout. Amphibian operations were unique and interesting and I really enjoyed it. I think that Bill was as much responsible for that as the aircraft and mission. He gave me such a thorough checkout that I really felt at home in the JRF. In some ways, the feeling of flying the Goose was similar to that of field carrier flying. It involved much of the old "seat of the pants" flying, where I was very close to nature (and the ground or water) most of the time. My JRF checkout went smoothly and Bill flew most of the training missions with me.

Usually, we flew up to the York River and set up a flight pattern where we made a dozen takeoffs and landings in a straight line, just working our way up the river. Our biggest problem was avoiding fishing and pleasure boats that shared the river, but the river was so wide, that it was not much of a problem. Bill showed me that the JRF was more of a split-power aircraft than it was a multi-engine aircraft. Most multi-engine aircraft can keep flying in the event one of the engines fails. The JRF could not. If one of the two engines failed, the airplane was coming down, even with the remaining engine at full power. This was true for practically any loading condition. In essence, we did not have the redundancy or reliability that a multi-engine aircraft typically had. Other than that, the JRF was a tough old machine built by the Grumman Iron Works, as Grumman Aircraft was known in those days. The weight and drag of the aircraft contributed to the lack of single-engine performance. In later years, some of the old JRFs had turbo-prop power plants installed, and could to get along on one engine if necessary.

The JRF had the hull of a boat and two outrigger floats, one on each wing, to prevent it from leaning or rolling too far over when on the water. The floats also added a potential problem. During amphibious landings, they could dig into the water too far and actually cartwheel the aircraft. One of the most critical performance regions for conventional flight is when the aircraft departs the earth as a high-speed ground machine and becomes airborne as a low-speed flying machine. The same is true for both land-based aircraft and seaplanes. Up to the point of takeoff, the amphibian has many characteristics of a speedboat. As it gains speed, the aircraft rises on the "step," which is a break in the hull that allows a plane or boat to cruise along on the top of the water with much lower drag. Once on the step, rudders maintain directional control. The pilot must apply opposite aileron to keep the wings level and not allow a float to drag or dig into the water. When he adds full-power for takeoff, the propeller wash showers the windshield with spray and reduces the pilot's visibility. As speed increases, the spray clears and forward visibility is no longer a problem.

NACA Langley retained the amphibian to provide support for operations at the remote rocket station on Wallops Island, Virginia. Wallops Station was located on the coast, north of Langley. The station, operated by Langley's Pilotless Aircraft Rocket Division (PARD), launched various rockets with aerodynamic models and airfoil test sections attached to instrumentation packages. High-speed photography allowed researchers to study airfoil flutter.

We operated the JRF between Langley AFB and a narrow creek or inlet just west of Wallops Island. The saltwater creek was subject to the ocean tides. Due to the variable depth of the water, there were specific hours that we could operate in and out of the creek. If the tide were low, the water was too shallow to permit operations. There was a steel mat located along one shore so we could taxi up to load and unload cargo. We would land, slow to taxi speed, and manually crank down the landing gear in preparation for exiting the water via the steel runway mat. When the water was moderately low and we had a priority trip, we had to taxi in a small circle where we knew the water was deeper and lower the landing gear in this area. Once the gear was fully down, we could actually taxi over hidden sandbars and not effect the landing gear. During extension, however, if the gear struck a sandbar, it was a mess for the maintenance and recovery people. In that case, we had to get divers to replace shear pins in the landing gear system before we could get ashore. Once we unloaded and reloaded, we taxied the aircraft back into the water. We then retracted the gear, set flaps for takeoff, and we were ready to go.

The operation was a little tricky, but fun. Bill Alford knew of the potential for trouble and he gave me a thorough checkout before he signed me off to carry passengers and cargo to the island. On my final sign-off flight, I demonstrated normal and emergency landings to Bill on the York River. After

NASA photo L-62-00296

The Wallops Island rocket test site was operated by Langley's Pilotless Aircraft Rocket Division (PARD). A narrow inlet on the west side of the facility served as a "landing strip" for Langley's JRF amphibian transport plane.

about an hour and a half of practice and training, Bill reached over and patted me on the shoulder. "Let's take it home Don," he said, "you are ready to handle the JRF." With the satisfaction of a job well done, I made the takeoff and turned south to head back to Langley. I had no sooner reached the treeline, at just several hundred foot altitude, when Bill cut an engine and I went to maximum power on the remaining engine and turned back to the river. I passed just over the trees and landed across the river. It was a good thing that no one was on the shore at this time. I did not have much altitude as we passed over the shoreline and plunked into the water. As we settled in, I turned to Bill with a dirty look. He just laughed and said, "You can't trust anyone or any machine." He also reminded me that I should always think about and look for a landing spot when flying the old split-power airplane. An engine failure meant coming down somewhere, soon. Bill was true to his word and signed me off that day, and I was always grateful for the thoroughness of his checkout.

I always enjoyed flying to Wallops in the JRF. At certain times of year, we shared the creek and our landing spot with oyster fishermen. When we approached the creek for landing, we would buzz low over their long, open boats and pull up into the downwind leg for landing. By the time we turned onto final approach, the fishermen were headed to shore. We flew our approach much like a carrier landing, with flaps full down and the aircraft just above stall speed in a low flat approach. The small creek that served as our landing area was bordered by marshlands and cattails. As the aircraft approached the edge of the marsh, I cut power to idle and made a full-stall landing on the water. The amphibian did not remain on the step and settled almost instantly. After we landed, cranked our gear down under water, and exited up the bank, the fishermen moved back to the middle of the creek to tong oysters again. When we finished unloading and prepared for the return trip, they cleared off to the side of the creek again. We did it all without any radio communications, and it worked well. They were able to get their oysters and we were able to deliver our cargo and passengers.

The JRF was used primarily for the Wallops support operation. The C-47 and the Electra were normally used for trips to Washington, D.C. and other areas on the East Coast. On one occasion, I had the opportunity to operate the JRF on ice and snow. Jim Whitten called me into the office one winter day and told me that he wanted me to take the JRF to fly Hartley Soulé to NAS Johnsville in Philadelphia. The C-47 and Lockheed 12A were both down for maintenance, and Hartley, who was "No. 2 Man" at Langley, had an urgent trip scheduled. I reminded Jim that there were no navigation radios on the JRF and a limited eight-channel VHF transceiver. This equipment was adequate for the visual flight rules that we operated under to Wallops Island, but not for other cross-country operations. Jim announced that he had checked the weather and that I shouldn't have any problems.

I checked the weather, too, and found there was about a foot of snow on the ground in the Philadelphia area, even though the skies were clear. Knowing how difficult it is to identify a field or town with snow cover, I drew up a sophisticated VFR track with time and distance from a known river bridge that I could identify in Philadelphia. The flight started well and the JRF was not too cold since the engine heat exchangers provided cabin heat on the way. Hartley was my only passenger.

As we approached Philadelphia from the south, the crew chief and I searched for the bridge we were using as a reference. We soon spotted it and I took up my heading for Johnsville. My earlier navigation preparations were critical since it was impossible to identify anything north of Philadelphia in that huge expanse of white ground. As I approached my time on heading, I initially could not identify anything that looked like Navy Johnsville. Finally, I spotted the round building that was their centrifuge facility. Even after I knew where the runway should be, I had trouble identifying the portion of it that had been plowed for landing. I slowed the JRF down just as I did on my water landings and made a slow touchdown on the snow-packed runway. The drifts along the sides were two to three feet high and a small exit was plowed at one end of the runway that led toward the parking area. The taxiway was sloped downward and it, too, was covered with ice and snow. Fortunately the parking area below was empty, because we slid down the taxiway and entered the parking area sideways.

Once the aircraft stopped sliding, we taxied and parked. Hartley did not seem too upset about our sliding arrival, but he was upset by the lack of heat on the final letdown and landing. With the engines reduced to idle power, heat production from the exhaust air heat exchangers was reduced to about zero. In the cockpit, we were busy with the problems of finding the runway, landing and getting to parking and we were not that aware of the cold, but Hartley had some pretty cold buns by the time he went into his Navy meeting. I heard later, after we had returned to Langley, that Hartley called Whitten and told him to never again use the JRF for the winter trips north again. I had tried to tell Jim that myself, but I didn't have enough influence to make him see reason. Sometimes, in the heat of the battle, you try to make do with too little.

I enjoyed support flying, even though it was not the same challenge as test flying. It was a nice change and there were many times that our test aircraft were down for instrumentation and maintenance and I was glad to be able to fly support. The fact that our support airplanes were new to me made it more challenging than it might have been otherwise. Since I flew types that I had not had the opportunity to operate in the Navy, it was new and interesting.

The JRF was strictly a Visual Flight Rules (VFR) operation. There were times in the summer when the visibility was limited due to haze and moisture in the Hampton Roads area. Sometimes during return flights to Langley from Wallops Island, we had limited visibility. In such cases, it was common for me to call Langley and request a "practice" ground-controlled approach (GCA) from the tip of Cape Charles to a landing at Langley. The JRF did not have a transponder, but gave a solid radar return, or "skin track," to the GCA unit. This worked extremely well, allowing me to find the airport easily on hazy, low visibility days. It helped coordinate me with other aircraft, including the higher-speed F-102 and later F-106 jets at Langley.

On one particular afternoon, the haze was worse than usual at sunset and visibility was very poor. I flew a fine precision GCA and, with a sigh of relief, saw the runway appear about half a mile ahead. As I prepared for landing, the tower told me to go around, and I soon found myself flying back into the "milk bowl" of haze and sun. I immediately called the GCA controller and inquired why they had sent me around. They said that on these so-called "practice GCAs," they did not give priority to the approaching aircraft and sometimes cleared other aircraft to cross the runway. The visibility had been so lousy that I hadn't seen any other aircraft. I acknowledged the controller's response, but I asked him for one of those "real GCAs." The next approach was uneventful and ended with a smooth landing.

Helicopters and VSTOL

Rotary-wing operations at Langley included both research flying and support operations. I first got up to speed on the helicopter support aspect. All the pilots in the office shared this duty, as many hours of support flying were required. Several helicopters were configured to carry and launch radio-controlled scale models of contemporary or future fighter aircraft. After release from the launch helicopter, the models were flown remotely from the ground.

NASA photo L-84196

The Sikorsky HRS-1 Chickasaw helicopter served as a launch platform for remotely piloted scale model aircraft. The models provided data on low-speed spin entry and spin recovery characteristics of various aircraft designs.

Researchers then evaluated low-speed spin entry and spin recovery characteristics. The program was of national importance and supplied free-flight model test and spin-tunnel data to the services and contractors who were conducting low-speed and spin tests on new fighters. The helicopters used for launching these models were the HRS-1 and H-19.

The models were launched from 8,000 feet and flown through planned test maneuvers. A recovery parachute was deployed by radio signal around 2,000 feet above ground level to recover the model. Flight test data was gathered in between these altitudes. Ground cameras with telescopic lenses provided long-range optical coverage. Sometimes special spoilers were attached to the model that caused it to enter an immediate spin following release from the helicopter. The spoiler was then released by radio signal and various recovery control inputs were evaluated to determine a good spin recovery technique and control for that particular aircraft.

Test results from the model experiments correlated well with full-scale test data. The models were very sophisticated and cost up to $20,000 or $30,000 each. We conducted the operations over Plum Tree Island, a marshy area east of Langley Field. The models were normally recovered without too much damage, but once in a great while, a model would land right in the middle of the six-foot wide saltwater creek that snaked through the marsh. When that happened, the majority of the electronics and instrumentation were completely fouled and required replacement. The ground-support people normally blamed that on the helicopter pilot because of his launch position, and we blamed it on the ground people. Of course, neither was really at fault. The model, under its recovery parachute, just drifted with the winds.

Research flying in helicopters was mostly limited to handling qualities and flight control evaluations. In a few years, a headquarters directive would change the emphasis of Langley's flight research to helicopter and VSTOL research, with many conventional aircraft test programs cancelled or transferred to the NASA Flight Research Center at Edwards Air Force Base, California.

When I was in the Navy, I stayed away from helicopter flying like the plague. I did not want to fly search and rescue helicopters from aircraft carriers looking for crazy fighter pilots. I wanted to be a crazy fighter pilot. Now it was different. Helicopter flying was a requirement of the job. As a NASA research pilot, I flew anything that came along or anything that I could work my way into. It was the life of a professional test pilot and there was always heavy competition for a slot in a new aircraft.

In forward flight, flying a helicopter is not much different from flying an airplane. The helicopter responds much like the airplane to stick and rudder inputs. The areas of difference that set rotary-wing flight apart from conventional flight are hover, and transition to and from forward flight. These areas require a great deal of coordination on the part of the pilot. It is a realm where little mistakes can result in big consequences. In addition to the normal control stick, called a "cyclic" in the helicopter, there is another stick called the "collective." This is the one that is new to the pilot previously trained in conventional aircraft. The collective controls engine power to the rotor and adjusts the angle of the rotor blades. This is very critical. The rotor speed must stay within an operation range of just a few hundred revolutions per minute (rpm). Exceeding maximum rpm can cause damage to the engine or transmission. Too low a rotor rpm can cause the helicopter to lose its hovering capability or even result in failure of the rotor blades.

To fly a helicopter, the pilot holds the collective stick in the left hand for power and rotor control and the cyclic stick, controlling pitch and roll, in the right hand. The pilot uses his feet on the rudders to control yaw. To complicate the task further, any change to the collective will induce some yaw input. During transition to forward flight from hover or from forward flight back to hover, inputs to the cyclic must continually be coordinated with inputs to the collective. The helicopter's movement and response to wind and gusty air provides an additional challenge. Noise and vibration of all the rotating equipment provides a chaotic background symphony. The early helicopters were powered by conventional piston-type engines. Modern helicopters, with turbine power, avoid most of the rotor rpm control problems. Also, the new electronic stabilization systems make them much easier to fly. I sometimes think that I was born a little too early, having to fly jets on straight deck aircraft carriers and relatively primitive helicopters with piston engines.

As I gained more rotary-wing flight time, I began to enjoy the experience. A person usually likes something that he is good at, and I found that true with helicopters. The more proficient I became in helicopters, the more I enjoyed flying them. It

NASA photo L-57-2263

The Hiller YH-32 Hornet employed ramjets mounted on the rotor tips for propulsion. Langley had two of these unusual little helicopters.

was a demanding, but satisfying, experience that paid off years later, after I had transferred to the Flight Research Center (FRC) in California. Because of my helicopter experience, I was assigned to the Lunar Landing Research Vehicle (LLRV) to provide flight test data for the astronauts that were going to make the moon landing in 1969. More on that later.

NASA Langley became more involved in helicopter and VSTOL research after 1960. A decision had been made to phase out the so-called high-speed jet flight testing at Langley and cancel or transfer the existing programs to FRC. Most of our jet programs were simply cancelled. I became more involved in the helicopter-VSTOL effort, as did all of the Langley pilots. I flew evaluations on some modified helicopters and qualitative flights on a few VSTOL aircraft. This included the Hiller YH-32 Hornet, which was powered by ramjets mounted on the tips of the rotors and the large, twin-engine HR2S-1 Mojave. I served as safety pilot on the HR2S while Jack Reeder put it through its paces. The helicopter's flight control system had been modified to remove the vehicle's natural stability by feeding negative control signals to the rotor system through the autopilot servos. Jack used it to determine just how poor a helicopter flight control system a pilot could handle and still accomplish a precision task. As Jack performed maneuvers over a designated ground track at about 50 feet altitude, I watched the control stick on my side (the right seat). The negative autopilot inputs, along with Jack's inputs, drove the system to 70 percent or 80 percent of full control very rapidly. When I finally ran out of nerve, I disengaged the test system and recovered using the normal helicopter flight controls. I sometimes think that the task was harder for the safety pilot, just sitting and watching as the machine tried to drive itself out of control.

Courtesy of Don Mallick

A Vertol Company representative congratulates Don Mallick, left, following his checkout flight in the VZ-2 experimental tilt-wing VSTOL aircraft at Langley in 1958. Mallick performed a qualititative evaluation flight in the aircraft.

I also had a chance to fly a qualitative flight in the Vertol VZ-2, a tilt wing VSTOL research aircraft driven by a small turbine engine and two propellers. The VZ-2 had a lot of cross-coupling between the roll and yaw controls and it was extremely difficult to fly. Generally, a qualitative test flight involves a pilot not assigned to a given aircraft program. He is given the opportunity to fly one or two flights just to take a quick look at the vehicle and glean as much information that he can from the limited flight exposure. He then writes a handling qualities evaluation report based on his experience. It was NASA policy then, as it is today, to allow pilots to fly qualitative evaluation flights whenever a test program can afford the time. It enables non-program test pilots to expand their experience by flying a variety of aircraft and provides additional data to program personnel.

After flying the VZ-2, I wasn't so sure I needed that experience. It took everything I had to keep from "dinging" or crashing the aircraft. During that time, I gained a lot more respect for Jack Reeder and Bob Champine, who were doing a lot of the VSTOL flight testing. I would have to say that the early VSTOL aircraft had extremely poor handling qualities and were very difficult to fly. I recall Jack returning from England after flying one of the early Kestrel VSTOL jet aircraft. "The British," said Jack, "are ahead of us again." He was very impressed with the airplane, all the more so as this was prior to the days of extensive artificial stabilization. The Harrier vectored the thrust of its turbofan engine through four louvered nozzles, two on each side of the fuselage. It was capable of vertical or short-field flight operations and was ideally suited for use on aircraft carriers.

As they said in the Navy about carrier pilots: "there were those who had crashed into the barriers and there were those who were going to crash into them, if you flew long enough." I suspect that test flying is the same. After flying long enough and being exposed to a number of different aircraft, an accident is probable, if not inevitable. I had my first (and fortunately only) major aircraft accident while flying a helicopter at Langley. Bob Champine and I were flying an HO3S-1, a relatively old helicopter. The HO3S-1 had a modified flight control system to evaluate flight envelope limits. Bob was flying in the rear seat of the tandem-seat helicopter and I was flying in the front, as safety pilot. Bob had the modified controls in the rear and I had the basic ship's controls in the front. Bob had just completed an evaluation and I took control of the helicopter and prepared to land on the runway. The HO3S was very sensitive to ground resonance, a very dangerous situation in some of the earlier helicopters. A harmonic vibration could occur during landing or liftoff when the helicopter weight was light on the wheels. The rotor created a force function and the response of the landing gear struts and tires served as a spring. A rapid, destructive vibration could develop in a matter of seconds. Our maintenance crews always kept our tire pressures within a pound of the prescribed pressure in order to counter this tendency. On this occasion, as I touched down, we encountered the dreaded ground resonance.

A pilot can respond to the situation with one of two options: push the collective firmly down and force the aircraft onto the ground or apply power and get back into the air. Either of these must be done quickly, and I chose the latter. I had a feeling that structural failure was imminent and snatched the

NASA photo L-83024

The Sikorsky HO3S Dragonfly had a modified flight control system to evaluate performance envelope limits. Its sensitivity to ground resonance caused its career to end in a catastrophic, but non-fatal, disaster.

helicopter back into the air with the collective. I added engine power and rotor rpm only to discover very quickly that I did not have enough right rudder power available to prevent a divergence in yaw that resulted in a roll and backward drift. Suddenly, I struck the runway with a resounding crash.

The last few seconds required for the helicopter to yaw and roll out of control seemed like an eternity. I can remember holding in full controls to counter the movement of the helicopter, but it was to no avail. The aircraft slowly diverged out of control until it struck the runway. One of the most terrifying moments in flight is when a pilot loses control and realizes that he is just along for the ride. The only analogy that I can think of is the feeling of sliding an automobile on an icy street. There is little to do except wait for the end of the ride or the crash. That was exactly how I felt. Another thing I noted was a lack of fear for my personal safety while this was happening. I was not saying to myself: "Boy, am I about to bust my ass," or something like that. I was actually thinking: "Boy, why doesn't this machine respond to my inputs and quit this crazy divergence."

We came to rest with the helicopter lying on its left side as the rotor beat itself to pieces on the runway. No fragments came into the crew compartment, and neither of us were hurt. The Plexiglass popped out of the cockpit windows and Bob and I exited unceremoniously, but quickly. I was one discouraged pilot, and had quite a dent in my pride. I had known pilots in the Navy who had broken up aircraft, some more than once, but I had always hoped to avoid doing something like that.

In those days, the accident board was quite informal. Both Bob and I described what we had observed, and our accounts were essentially the same. I wrote up a short statement describing the accident as I saw it. The accident board reviewed the flight data and confirmed that ground resonance had occurred. I felt it was obvious that I had not handled it well. There was no reprimand or disciplinary action, but for a week or so I waited for the other shoe to drop. One afternoon, Hartley Soulé came by the operations office and talked to me about the accident. I knew that Jack had already filled him in about it. I told Hartley that I was sorry about breaking up the machine. "Hell, Don," he said, "that old helicopter was at the end of its useful life and I was wondering what we were going to do with it." It was his way of saying "don't sweat it." He was glad that no one got hurt. It was still a blow to my pride, but I'm not sure that I could have changed the situation. If I had the chance to turn back the clock, I would have selected the other option of slamming the helicopter on the ground when the resonance started. Some people thought that part of the rotor hub had failed prior to my pulling the helicopter back into the air, but that was difficult to prove from the remaining debris. I was sad and suffered a damaged ego, but never considered giving up flight testing. I loved flying and just hoped that I didn't have to put up with too many of these situations in my career.

4

Jet Research at Langley

NASA Langley Jet Operations

When I first interviewed for a flying position at Langley in the spring of 1957, the lab was involved not only with helicopter and support flying. There also were a number of interesting research programs related to the handling qualities and control systems of various jet aircraft. NACA (later NASA) aeronautics research was divided among several laboratories. At this time, the Lewis Flight Propulsion Laboratory in Cleveland, Ohio was involved primarily is aircraft propulsion research. The Ames Aeronautical Laboratory at Moffett Field, California conducted jet programs and some VSTOL aircraft operations. The High-Speed Flight Station at Edwards Air Force Base, California flew jet and rocket-powered research aircraft.

Jet aircraft in the "stable" at Langley included a number of Navy fighters: F2H-1, F9F-2, F9F-6, and F11F. The U.S. Air Force fighters on hand were the F-100C, F-86D, F-101A, and the exotic XF-88, a prototype fighter that had been converted with a turbo-prop powerplant in the nose to evaluate supersonic propeller development. Most of the aircraft were used to study flight control systems and flying qualities.

My first jet test program was on the F2H-1. It was very similar to the F2H-2 that I had flown in the Navy. One of the primary differences was that the F2H-1 at Langley was much lighter, but had the same engine thrust. This made it seem like a little hotrod. The test program involved a "g-limiter," a system that could be adjusted to limit the acceleration loads that the pilot could impose on the aircraft in turns or pull-ups. The original aircraft "g-envelope" had a maximum limit of 7.5 times the force of gravity (g). The aircraft's performance was such that the pilot could actually pull enough g's over the limit to damage or break the wings. At the time, the g-limiter on the aircraft was a mechanical friction brake with a geared system that actually locked up the elevator when a preset g-limit was achieved. The aircraft was equipped with accelerometers that were summed and averaged, and pitch rate signals to prevent overshooting the limit. Amplifiers fed this information to the braking unit installed on the back of the aircraft. On initial developmental testing, the onset of acceleration or loading used was moderate. This allowed the pilot to monitor the acceleration build-up to the limit level. In the event the brake failed, the pilot could ease off on the stick input to avoid exceeding the aircraft's structural limits.

After I proved it would work with moderate inputs, I used snap inputs to evaluate the overall function of the system. I would pull full-aft stick as rapidly

Jet Research at Langley

NASA photo L-74183

Mallick's first jet test program at Langley involved the McDonnell F2H-1 Banshee. It was modified to limit the acceleration loads the pilot could impose on the airframe while maneuvering.

as possible and rely on the elevator brake to protect the aircraft from structural failure. These flights were always a challenge and added risk to an otherwise routine test. The program went extremely well. As a result of the large stick inputs required, the maneuvers were nothing short of acrobatic. I made vertical pull-ups and recovered into loops or Immelmans, as the situation required. I became very proficient flying the unusual attitudes that resulted from gathering the required data for this program. There was not a straight-winged jet flying at that time that could outperform that F2H-1 with its lightweight airframe and powerful engines. The F2H-1 was also used as a chase aircraft.

Simulation

For decades, the Navy and Air Force have used simulators as training devices to prepare pilots and keep them proficient. There have always been arguments about just how realistic simulators are and how much they help. At Langley, simulators were used as research tools, providing a great deal of useable data. If you ask most pilots, probably 99-percent of them would rather fly the real thing than a simulator, but they also recognize that simulators serve a vital function. As a junior pilot, one of the "new boys" at Langley, I received my share of simulator flying, while some of the older, more senior pilots flew the real thing. I admit that this provided excellent background and

The Smell of Kerosene

experience required for a research pilot, and I was fortunate to be exposed to it at this point in my new career.

One of my first research simulation assignments involved the vertical motion simulator (VMS). The VMS consisted of a pilot's seat mounted on rails in one corner of the big aircraft hangar at Langley. The seat had motion in a vertical axis of plus-or-minus 30 feet. It also had a rotation in the pitch axis of plus-or-minus eight degrees. The chair, on specially designed tracks, was propelled by a hydraulic mechanism from a Navy five-inch gun mount. The seat had very high vertical acceleration capability. There were large buffer springs at the top and bottom of the track to repel the chair in the event the pilot lost control. If he lost control, the seat (and pilot!) slammed hard against the spring stops before the motion of the simulator damped out and stopped. The suffering incurred by this event encouraged the pilot to keep the machine under control.

My task in the simulator involved discovering control limits and evaluating pilot inputs. Part of it included a tracking task. A target ball moved vertically up and down in front of the pilot's position. It was driven by a variable input that could be changed in order to prevent me from memorizing a pattern of movement. A gunsight with a projected image was mounted behind me, and my task was to keep the gun sight crosshair on the moving ball. An error integrator measured my performance in tracking the target. The evaluation was designed to determine a pilot's ability to visually track a target when his aircraft had very poor stability and handling characteristics, even to the point of static instability. "Statically unstable" means that the aircraft would diverge and crash if the pilot failed to keep it under control. I was already extremely busy just keeping the simulated aircraft from diverging. The tracking task increased my workload. High stick-forces assisted me in making small precision inputs. When the simulator reached the boundary of my capability, the stick force had reached 800 pounds per g acceleration. This compares with about eight pounds per g on an average fighter aircraft and perhaps 20 pounds per g on a transport or bomber aircraft.

The program engineer on the vertical motion simulator was a likable fellow named Porter Brown. Although I enjoyed working with old "Porty" on this program, he sometimes worried me. I was flying at levels of negative stability that would never be accepted in any aircraft. In fact, the simulator was so sensitive at reduced levels of stability that I could fly a condition on one day and not fly it the next, simply due to differences in my personal responses and capabilities on a particular day. I can recall that Porter was frustrated with me on days when I could not control the simulator. He said that I had flown the same condition a day or two before, so I should be able to duplicate my previous performance. In the simulator, I felt as if I was balanced on the top of a moving ball and at any instant I might fall off. I was operating on the edge of controllability.

Porty and I would sometimes "knock heads" about his program, and he would accuse me of being a "ham fisted" pilot. I would then accuse him of writing his reports and results before the data was in. I think it was just the nature of the program. In an effort to define the boundaries of a pilot's capability, we were trying to do something that was not always repeatable. Porty was highly motivated and he had been instru-

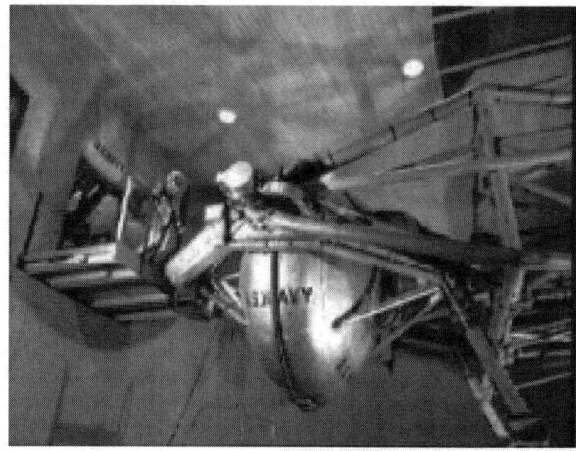

NASA photo G-60-2461

NASA employed the Navy centrifuge at Naval Air Station Johnsville to learn how astronauts would be affected by high acceleration forces during launch and re-entry from spaceflight. Langley research pilots also conducted test programs in the centrifuge.

mental in designing the VMS. In fact, knowing Porty, I think he designed the wild, out-of-control response of the vertical seat just to give the evaluating pilots more incentive to retain control. When the machine "departed," it was a thrilling, tooth-rattling, eye-shaking experience with two to three rapid cycles against the upper and lower spring stops that almost snapped the pilot out of his seat belt and shoulder harness. The seat then came to a stop at the neutral position and returned to what we called "IC" (Initial Condition). Porter eventually left to join the Space Task Group when it was formed in 1958.

Another simulator available to us was the Navy centrifuge at NAS Johnsville in Philadelphia. The Navy used it for evaluations of pilot capabilities and limitations while operating under high g-forces such as those experienced in fighter aircraft. With the advent of the space program in 1958, concerns arose as to how the astronauts were going to perform tasks while under high acceleration forces during launch and re-entry into the atmosphere. The research pilots at Langley were involved in a number of programs on the centrifuge to evaluate a human's capability under high g-forces.

The centrifuge was located in a large circular building. A central mechanism drove a large arm. A cockpit gondola, gimbaled on two axes, was attached to the end of the arm. The pilot's display and seat support could be changed in order to evaluate a number of configurations. A computer allowed the pilot to "fly" a mission from the gondola simulator and recorded the accelerations that resulted from his control inputs. There were also "canned" programs where the pilots were exposed to various g-level plateaus. During exposure to various accelerations, the pilot was required to perform functions in the cockpit such as switch and actuator motions to determine the effects on his capabilities. His performance, timing, and accuracy were carefully evaluated. I participated in a number of centrifuge programs over the years at Langley and Johnsville. Although they were interesting, they were physically demanding. There were days when I was in the gondola for more than four hours at a stretch. I never worried about where the "nightlife" was after dinner in Johnsville or Philadelphia, as I was more than ready to get some rest in order to be ready for the next day "on the wheel."

The Langley pilots shared the centrifuge programs. Each of us took a program now and then. Centrifuge work took us away from our regular test flying, but it had to be done. It was not very popular, however. During the centrifuge programs, service pilots were occasionally invited to participate. Once, a Marine Corps major came in for a day or two to fly the simulation. He was a personable fellow and seemed to have boundless energy and exuberance. He had recently set a cross-country speed record. Even though I only knew him for a couple of days at the centrifuge, I could not help but be impressed. I wasn't surprised, some months later, when he was selected as one of the original seven astronauts. That pilot was John Glenn.

In 1958, the NACA became the National Aeronautics and Space Administration (NASA). Things were hectic at Langley as people sought to find where they fit into the new organization. There were some pretty wild proposals going around, initially, as the engineers and scientists worked to put a program together to get a man into space. There were also some pretty unusual simulation projects started and, fortunately, not all were completed.

The one that I recall best (because I was assigned to it) was a real "spook" of a program. The plan involved evaluation of a pilot under conditions of simulated "zero g" (i.e. floating weightless). A large spherical steel tank was welded together and assembled at the rear of one of the engineering buildings. The windowless tank was about 30-feet in diameter. The plan called for a pilot's seat in the center of the tank and an air supply for the test subject through an oxygen mask and hose. A communications link would connect the occupant to an outside monitoring station. The interior of the tank was to be completely dark and filled with water! The purpose of the test was to see if the subject pilot could determine his orientation under zero-g (simulated by floating in water) as his seat was moved to different positions in the center of the tank.

This project developed quite quickly. The test pilots in our office were called in, almost as an afterthought, when it was decided that we would be good subjects to conduct zero-g research. I had serious concerns regarding the safety of this large steel ball full of water. How was a test subject to get out if there was a problem with the breathing or communications system? I was informed that the designers had consulted an "expert," a NASA engineer who was also a licensed civilian deep-sea diver. This "consultant" had approved a

Mallick flew a handling qualities evaluation program in this JF-100C. The aircraft was modified for variable stability performance.

single hatch on the bottom of the tank that could be opened in emergency situations. On opening, the tank would empty in about three minutes. The whole design philosophy of the tank seemed poor, as far as safety was concerned. This was probably a result of the immediate hysteria that was running through the agency with regards to the new "space race."

Another test pilot and I were recruited into the project. We started to take SCUBA diving lessons in the indoor pool at Langley in preparation for the simulation. According to the plan, the consultant engineer with diving experience was going to be the initial test subject for the tank. Once he had pronounced it safe, the test pilots would have their turn. As it turned out, on the first or second trial in the ball, something failed in the breathing system and the diver almost drowned. Following an investigation, the whole project was cancelled. I think I had an extra scotch and soda that night to celebrate. First, I was glad the diver survived, but almost as important, I was glad that this particular program was scrapped. I never had a second thought about continuing as a test pilot after the helicopter crash, but I was considering thoughts of giving it up when faced with assignments like the water tank. With a little time, cooler heads prevailed at NASA as organizational responsibilities were ironed out, and the scientists and managers settled down to the task at hand.

In the fall of 1958, I prepared to fly Langley's JF-100C Super Sabre jet fighter. The JF-100C had a modified flight control system that could vary the stability of the aircraft, and I planned to gather some handling qualities data. Nearby Langley AFB had an F-100 simulator. After reading the handbook and studying the aircraft systems, I scheduled several simulator flights for practice prior to flying the actual aircraft. Using every means available to prepare a pilot for flying a new aircraft was central to the philosophy NASA's flight test organization. I'll never forget that after my first ride in the simulator, I told the Air Force sergeant operating it that it must need some adjustment. The longitudinal characteristics of the F-100 on final approach were poor. The stick felt lumpy. The control response was sensitive and it was difficult not to oscillate the aircraft around the pitch axis during the final approach. The sergeant seemed a little let down and said that none of his Air Force pilots had complained, but he would check out all of the parameters to make sure that nothing had slipped or changed in the simulator's programming.

A few days later, I made my first flight in the JF-100C. The longitudinal control felt exactly the same in the aircraft as in the simulator. The additional visual and physical cues that were available in actual flight helped a great deal. It was not as difficult a task flying the aircraft as it was the simulator, but the longitudinal flight control on the JF-100C on final approach left something to be desired. I made a point of telling the simulator operator the very next day that his simulator was fine. It definitely resembled the actual aircraft. Characteristics such as this are more difficult to fly on fixed-base simulators where there are no motion cues and limited visual cues. In the aircraft, the pilot has some g-feedback with the stick input. This gives him more information to work with and helps prevent the over-controlling.

My experience with the JF-100C provided an important lesson. When a pilot evaluates an aircraft that has difficult flying qualities, the problem areas are usually apparent even if it takes a little time to sort out in detail exactly what is causing the problem. After the same pilot experiences these problems repeatedly or continuously, his mental "computer" (i.e. his mind, reflexes and responses) begins to automatically compensate for the shortcomings of the aircraft. He is no longer bothered by the problem. Of course, there are limits to the pilot's capabilities. His mind (the "computer") and his servo actuators (his nerves, muscle, and body structure) have definite limits as to what they can handle. In World War II, the P-47 Thunderbolt was an excellent fighter and the majority of the P-47 pilots swore by their aircraft. Actually, the aircraft's static stability was marginal and its maneuvering stability was negative a great deal of the time. This means that while pulling acceleration or g's in high-speed turns, the pilots would have to push on the stick rather than pull normally. Considering all of this, the P-47 pilot liked his aircraft and was able to perform admirably with it.

Does this mean that aircraft designers and builders need not consider the handling qualities or flying capabilities of their aircraft? Certainly, not. Flying qualities are very important. The pilot's "computer" and "servo mechanism"

NASA photo L-58-1840

This Grumman F9F-2 Panther had a modified flight control system that allowed the pilot to evaluate a wide variety of test conditions. Mallick used it to establish a minimal (yet satisfactory) handling qualities envelope for the Navy's operational Panther jets.

is a wonderful system designed by God and nature, but it has limitations and saturation levels. If the pilot's capabilities are largely occupied with keeping the aircraft flying, he will likely become saturated during demanding maneuvers or emergency situations.

Another area where poor handling qualities become apparent in an aircraft is during the familiarization stages of flying. The pilot is in the process of learning and compensating for the aircraft. If he encounters an especially difficult condition in flight, the combination the aircraft's flying qualities and the pilot's experience can wind up as an incident, accident, or at least a good scare. In my opinion, there is no substitute for good flying qualities in an aircraft.

Other Interesting Jet Programs at Langley

At Langley, we had a Grumman F9F-2 Panther jet with a very unusual flight control system. A side-arm controller could be adjusted for force input, displacement input, or a combination of the two. In addition, the control system of the aircraft could be set up for rate response or acceleration response. The modified controls allowed us to evaluate a wide variety of test points and conditions to aid development of improved flight control systems for future

NASA photo L-57-2254

Mallick flew a qualititative evaluation of the Grumman F11F-1 Tiger. He found it had excellent handling qualities, but only marginal engine performance for operational use.

fighters. To this end, we experimented with air-to-air tracking of a target aircraft, usually the F2H-1 discussed earlier. The pilot's measured tracking performance, along with his qualitative evaluation, provided the desired information.

The aircraft's center of gravity was also varied for the evaluation. The Panther's fuel transfer system was modified and the pilot had the capability of transferring fuel aft following takeoff to cause the aircraft to become statically and dynamically unstable to the point that it would lose control if not held steady by the pilot. This presented the pilot with a difficult task in tracking a target aircraft while flying an unstable aircraft. The purpose of the center of gravity control was to enlarge the range of the evaluation envelope. The pilot transferred fuel forward prior to landing, to return the aircraft to a safe flying condition, and then land using the conventional flight control system.

Technicians also mounted a small steel container far back on the lower fuselage and loaded it with 250 pounds of Number 7 lead shot. This shifted the center of gravity aft near the unstable condition but within a safe margin for takeoff and landing. In an emergency the pilot could activate two explosive bolts to jettison the contents of the box, allowing an immediate forward shift of the aircraft's center of gravity. The small size lead shot was selected to reduce the threat to anybody on the ground, in the remote chance that it fell on them after jettison.

During maneuvers and tracking, there were times that the pilot had to push forward on the stick while maintaining a 4g tracking curve on the target aircraft. We often flew with very poor handling qualities while still attaining satisfactory target tracking. These conditions, however, were unsatisfactory for an operational aircraft. One false move, or relaxation of concentration, could result in a pitch-up or loss of control.

The program to establish the minimums for a satisfactory handling quality envelope was interesting. It was the first time that I had ever flown an unstable aircraft. Visiting U.S. Navy and Air Force test pilots also flew the airplane to evaluate the experimental flight control system.

Langley had a Grumman F11F-1 Tiger in those days. It represented a relatively current Navy fleet aircraft and it was a pleasure to fly. The airplane's handling qualities and responses were excellent. After a long career of flying many different fighters, I cannot think of any that would surpass the Tiger. The aircraft was most impressive in the landing configuration, where it was rock steady and easy to fly. The up-and-away characteristics were equally nice because the stick force and displacement-per-g were ideal. The only problem with the aircraft, one that had occurred on other fighters of this period, was that the powerplant-airframe match was not good. The Tiger had a J65 engine with afterburner, but its performance in military power was inadequate for a fighter. The engine's performance in afterburner, while acceptable, consumed fuel so rapidly that it was not practical for operational purposes. This is one of the reasons the F11F-1 did not remain in fleet service very long. A good carrier-based fighter must have a reasonable thrust-to-weight ratio and range. A few were modified with the higher-thrust J79 engine. Although performance improved, the range and endurance did not because there was limited fuel storage capacity in this beautiful little airplane.

In 1958, NASA Langley also operated a Vought F8U-1 Crusader. Like nearly all NASA test aircraft, it was completely instrumented to provide flight test data. Bob Champine served as the research pilot for the program. The Navy had suffered two or more incidents where their pilots had pulled the wings of similar F8Us during high-g maneuvers. The fleet did not have any appropriately instrumented Crusader's in their inventory and needed information quickly. Navy officials asked NASA for data to help determine why the wings were coming off. Langley management agreed because support of the military services was very much within their charter. NASA received a great deal of support from the military, including loaned aircraft to use in flight research programs.

The F8U was unique in that it had a variable incidence wing, one that tilted up and down from the fuselage as flight conditions required. For takeoff and landing where a lot of lift was required, the wing was raised about 10-degrees to give a higher wing angle of attack without increasing the fuselage angle. The reason for the design was to avoid having a long nose wheel strut

NASA photo L-57-2099

Bob Champine served as project pilot for this fully instrumented Vought F8U-1 Crusader. Mallick observed the flights and reviewed the data with Champine and the Langley engineers. The project established a performance envelope for the aircraft.

like the F7U that created problems in carrier operation. In any case, the accidents that had occurred with the F8Us resulted in the complete wing assembly leaving the aircraft. As a result, the variable incidence wing and its mechanism were suspect.

The F8U-1 had an actuator on the left side of the fuselage that extended the wing to the takeoff and landing position and retracted it to the cruise, or high-speed, position. The wing failures had occurred in the cruise high-speed condition under high g-forces. The right side of the wing had a "follower" strut, but the main wing locking mechanism was on the left-side actuator. NASA quickly added a fuselage-mounted, high-speed motion picture camera on the left side of the aircraft. It pointed at the junction between the wing and fuselage, which was normally very streamlined and smooth while the wing was retracted in the high-speed cruise configuration.

Christopher Kraft served as the project engineer on the F8U-1. Chris was a well-respected engineer who joined the Space Task Group in 1958 and later made quite a name for himself as a Flight Director for manned space missions. With the F8U-1, Chris set up a series of flights, increasing the g's on the aircraft in gradual steps. The strain gauge data took some time to process. Therefore, engineers used the camera data to clear the aircraft to a higher g-level on the next test flight. There was an air of urgency with regard to helping the Navy find the problem and correct it to restore the fleet aircraft to full readiness.

I did not fly in the F8U-1 program, but Jack Reeder directed me to follow the program for my professional growth. I observed the test flights and reviewed the data with the project pilot and engineers. When Bob flew a test flight that included 4g turns, the camera data showed an opening between the wing and fuselage on the left side. Although it was less than one half inch, it was a definite and unexpected movement. Some of the engineers thought it might be a normal movement, similar to a wing tip bending upward under load, but no one knew for sure. The big question involved whether to ground the aircraft for a day or two to mount another camera on the right side of the fuselage to confirm the motion. Chris was anxious to complete the program. As leader of the engineering group, he proposed flying to higher g-levels without the other camera. Bob, as project pilot, preferred a more conservative approach. He seemed reluctant, however, to disagree with the engineers. At the postflight analysis meeting, the consensus seemed to sway toward another flight the following day.

I was a test pilot in training with a little over a year's experience, but the idea of flying again without that starboard camera installed seemed risky to me. I couldn't sit quietly. Patiently, I waited for a break in the discussion. "I wouldn't fly that aircraft to a higher g tomorrow," I said, "without the right-side camera." There was a silence in the room as the "old heads" all turned to see where this comment came from. I could sense what they were probably thinking: "What in the hell does this new guy know?" In any case, another pilot's input was enough to sway the decision. The aircraft was grounded for a few days to mount the camera. The U.S. Navy had to wait.

I suspect the Navy would have gladly waited if they had known the facts. I think Chris Craft was a little mad at me at that time. I wasn't mad at him, I just thought that it shouldn't fly without the camera. Bob Champine flew in two or three days with the new camera added and he repeated the 4g maneuvers. When the film was analyzed, a few people caught their breath. The right side of the wing at the fuselage was rising up over two inches, indicating a serious unexpected wing twist. NASA quickly notified the Navy and the aircraft performance envelope was limited to 4g's. Further investigation showed that the hold-down mechanism of the wing was inadequate on the right side and the follower actuator was not contributing much, other than beauty. A modification was in order.

This experience provided me with a valuable lesson. It is always nice to be right, and I certainly gained some credibility from the incident. The lesson learned, however, was not to let peer pressure or politics keep me silent when it came to flight safety. All through my career as a test pilot, I witnessed similar situations where there was tremendous pressure for one reason or another to continue when I felt it was the wrong thing to do so. I always tried to make my decision on the side of safety and conservatism, and take an extra day if it was required. I truly believe that is one of the reasons this old test pilot is alive today. As they told us years ago in the Navy: "There are old pilots and there are bold pilots but there aren't many old bold pilots around." I continued to exercise my philosophy when I became involved with flight test management. Although I generally gave the project pilot wide latitude for responsibility and decision-making, there were times that I would override his decision on the basis of safety if I felt that it was appropriate.

Langley Joins the Space Race

The nucleus of NASA was the "old" NACA. It was a government research organization that was truly not a political animal. The administrators of the NACA over the years were scientists who had worked their way to the top of the organization. The government funded "the Committee" (as it was called in the early years), but its budget and visibility were relatively small. There was very little government pressure on the organization, a desirable situation for a research organization. It provided an atmosphere conducive to experimentation and invention.

When I entered the NACA in June 1957, I had no idea that I was entering at the end of one era and getting ready to begin a new one. The challenge of aeronautical research was tremendous and it was what I had envisioned for my career when I became a test pilot. As a NACA research pilot, I was in a very enviable position. I experienced the challenge of flight test programs found nowhere else in this country.

When the Soviets shot their Sputnik into orbit in October 1957, speculation ran rampant as people tried to guess how the U.S. would respond to this achievement. The various military services lobbied hard, vying for the responsibility of conducting the U.S. space effort. Some thought the NACA a logical agency to direct such a program since the Committee had a core of fine scientists and research facilities. President Eisenhower and his administration charged the NACA with responsibility for space as well as aeronautics. In the reorganization, it became the National Aeronautics and Space Administration (NASA).

Langley was selected as the initial home of Project Mercury, the U.S. manned space effort. A large group of scientists were recruited and transferred from the aeronautics side of the house to the space side. It was obvious that space research was going to dominate the agency, due mainly to the scope of the task. The initial effort was referred to as the Space Task Group and it was located in some of the old NACA buildings on the southeast side of Langley Field. It is not difficult to imagine the confusion as engineers and scientists of the original aeronautics groups made their decision to switch over to space research. They entered a realm of endeavor that was outside their previous experience. The organizational and planning problems were tremendous. It was as if someone threw a switch and said, "You are now working on spaceflight research. Where are we going and how do we get there?"

I give a lot of credit to those early people that got the organization under way. They "picked up the ball and ran with it." Ultimately, they designed and tested the Mercury spacecraft, trained the original seven astronauts, and developed a tracking and ground instrumentation system to monitor each mission. As efforts moved from basic research to project support, Langley Memorial Aeronautical Laboratory was renamed Langley Research Center.

For myself, there was no decision to make. As a research pilot, I was exactly where I wanted to be. Had I been working strictly in engineering, I probably would have switched over to the Space Task Group. Fortunately there were a number of top-notch scientists and engineers that remained in aeronautics. Those of us that "stayed behind" observed the birth of the Space Task Group from a close position, but with little involvement. Some of the early proposals ran the gamut from the outrageous to the macabre. Some scientists suggested that astronauts might have to have their legs amputated in order to sustain the acceleration of launch and reentry into the Earth's atmosphere. This, they said, would prevent the loss of consciousness that occurred when blood pooled in the legs. This was just one of many "incredible" concepts that surfaced during this frantic phase.

NASA seemed like a giant "think tank" in the early days. All NASA scientists and engineers were encouraged to focus all of their thoughts toward how to put a man into space and return him safely to the Earth. Along with the wilder ideas noted, many more practical notions were put forth. Scientists from different areas within NASA came up with similar proposals. A scientist named Maxime Faget proposed a plan that led to the Mercury, Gemini, and Apollo spacecraft. Max was small in stature, but a mental giant. He proposed using a capsule-type vehicle as a manned spacecraft. The capsule approach was the quickest method to get a man into space. It required a parachute landing in the ocean and all of the support services of the Navy for recovering the capsule. This added some risks and problems, but overall, it was the best way to go in a time-critical effort.

I am cautious when trying to speak for others, but in this case I think that I can speak for my colleagues in the Langley pilot's office. We felt that the space effort, as it was emerging, involved a rush job that was more political than scientific. It seemed that beating the Soviets with a man into space was the primary goal in this new "Space Race." It appeared irrelevant whether or not the man in the capsule was to be involved with flying the mission or just a passenger. The objective was to get him there and get him back alive. I have always been curious as to what the first space ship would have looked like had we not rushed ahead so quickly. I suspect that it would have been like a small shuttle or perhaps the X-20 Dyna-Soar program that was ultimately cancelled. In any case, I feel that there would have been more active participation by the pilot-astronauts.

As NASA began its march toward space, its budget increased substantially. The NACA had survived for years on a relatively stable budget. On occasion, extra money was provided for special programs like the X-15, but it was essentially a moderate budget. This changed when NASA was formed. An appointee of the President now led the agency that had been guided for years by scientists. The new administrator was more of a politician than a scientist. I'm not criticizing this. I think it was a necessary evil to help NASA work with Congress and the administration in order to preserve the budget. I feel that there were times, however, when the technical side of the house suffered because of politics. The public visibility of NASA worldwide turned into a political tool for the administration. The President gained visibility whenever he talked by phone with astronauts during space missions. Also, space launches enhanced the status of this country in the world community.

Although the NASA budget declined during the Apollo program, the nation's economy was booming and there was strong political and public support for the program. Following Apollo, things began to change. New administrations and a weaker national economy required NASA to provide more justification and fight for its budget. Unfortunately, NASA entered an "oversell mode" and promised much more than they realistically could deliver for a given budget and time. This frequently put the Agency behind schedule. NASA tried to accomplish missions to satisfy a "political" pressure, with scientific considerations receiving lower priorities. NASA has always stated that flight safety is the top priority, but over-optimistic schedules in the political limelight made safe operations more difficult.

Langley Support and Test Flying

As expected, support flight operations at Langley grew rapidly with the conversion of the NACA to NASA. We were no longer just involved with flying to Wallops Island and Headquarters. We now flew to Cape Canaveral, Florida; Huntsville, Alabama; and number of other cities that were involved in the space effort. We carried engineers and managers from Headquarters to various assignments and meetings. Langley acted as the primary logistics flight support for NASA Headquarters. Lewis Research Center near Cleveland, Ohio also assisted with this mission. NASA procured a surplus Air Force T-29B aircraft and converted it into the NASA executive aircraft. Unfortunately, the T-29B had gained an excessive amount of weight with its interior conversion and this somewhat reduced its takeoff performance. We watched our operating weight closely to maintain a safe single-engine performance capability.

I experienced a great deal of instrument flying in the T-29B on the East Coast in the winter. I can recall a number of instrument approaches into Langley AFB in the early morning hours where the fog was thick and the visibility practically zero. It is amazing just how alert a pilot can be at 3:00a.m., after a long day of flying, when he is trying to find a runway in such conditions.

Many times the runway was barely visible prior to touchdown and the only saving grace was the high-intensity center line lights that came out of the fog at the last minute to assure me that I was somewhere near the runway. Shortly thereafter, the cross bar light arrangement gave me enough information to make a few small bank corrections to centerline. I made my touchdown on instinct because I could barely see the runway. Rollout to a stop wasn't too difficult since the landing and taxi lights illuminated the white centerline painted on the runway.

Finding my way back to the NASA taxi strip was sometimes more difficult as the white line leading from the runway over to Langley was much smaller and more difficult to see than the runway center line. I can remember the relief that came over the crew once the aircraft was stopped and chocked in front of a hangar that we couldn't see. I knew it was there because of the taxi lines on the ground. Flying like this built character and gave me a feeling of confidence, knowing that I could handle that type of weather. I would be lying, however, if I said that I enjoyed it. I much preferred a little more visibility, especially close to the ground. The commercial equivalent of the T-29B was the Convair CV-240. The Langley pilots were getting a lot of "airline" type flying. Although it was not as interesting as test flying, there were times when it challenged our abilities. That, too, is satisfying.

During conversion of the T-29 to an executive aircraft, Headquarters refused us the money needed to add a water separator to the air-conditioning system. Butler Aviation, the company making the conversion, had emphatically insisted that we add one because the military air-conditioning system was completely unacceptable for commercial planes. Expansion-type turbines that provided cooling also produced a tremendous amount of water. The military water traps were not efficient enough to prevent heavy fogging in the cabin. Butler had earlier experience with the newly added headliners and interiors falling loose due to excess water in the cabin. Our Headquarters people were not convinced by these arguments and the aircraft went into operation without the added water separator.

About three weeks into the operation, another pilot and I were assigned to fly a group of people from Headquarters to Cape Canaveral. We departed out of Washington on a very warm, muggy night. As we climbed out on our instrument departure, the crew chief and steward came forward laughing and told us to look back in the cabin when we had the opportunity. The cabin was obscured by fog. There were small blue footlights along the aisle, and the whole cabin had a spooky, blue, swirling atmosphere. The following Monday, Headquarters informed Langley flight operations that money was now available for the water separator. Additionally, money was provided for some special muffler kits for the engine exhausts over the wing. It was very difficult to carry on a conversation in the cabin, even during cruise operations, with the original round exhaust stacks that exited over the top of the wing. With the modifications, the T-29B became an acceptable executive aircraft for NASA and was used for some years before it was replaced by a Grumman Gulfstream I aircraft.

Langley also conducted photographic operations in support of the space effort. We had a T-33 aircraft for photography and chase during test missions. This aircraft had two unusual photo applications for the space effort. The Pilotless Aircraft Rocket Division (PARD) conducted flutter tests on small airfoil sections out of Wallops Island. Technicians assembled solid-fueled rockets with airfoil test sections mounted on the rocket bodies and special data cameras to record its performance in flight. The rockets were fired over our Atlantic Ocean test range. High-speed data cameras recorded the flutter or destruction of the test surfaces. The test vehicle then parachuted into the ocean for recovery by boat or helicopter.

PARD had been doing this for some years, and they were proficient in the operation of these smaller rockets. It was natural for them to conduct some of the tests required for the space program. One test involved a five-stage solid rocket that was fired into space on three stages. The final two stages propelled the vehicle back into the Earth's atmosphere to gather data on temperatures achieved at spacecraft reentry speeds. The rocket had a spin-stabilization system, but was otherwise unguided. Some data was telemetered back to the station, but the PARD scientists were very interested in having pictures of the re-entry in order to study ionization of the air around the rocket at extremely high speeds.

There were several long-range, ground-based cameras located south of Wallops on the east shore of Cape Charles. These cameras were at the mercy of the weather as clouds and smoke could hinder their data-gathering ability. PARD therefore requested that Langley provide a photo plane to fly at 35,000 feet in the vicinity of the reentry path to obtain some pictures. The task was assigned to me, and I set about the planning with the engineers of PARD. One of the most difficult things about the operation was that, due to lack of exact guidance, the rocket engineers only had a predicted area of reentry. They said they would take whatever we could get for them.

The PARD people also gave me the reentry time, which helped my planning. I selected a ground speed for my aircraft and worked out a flight plan to place me in a specific spot over the ocean at the predicted re-entry time. I had to use the very latest upper-level wind information available, and select a heading and indicated airspeed to assure my arrival at the desired location. I also had to control the final countdown for the missile launch so that I departed my VOR fix at a specific time. In all, it was a chancy operation. There was no guarantee that I would see the missile reenter, but PARD wanted to try it. Since I flew a visual omni-range (VOR) radial out of Salisbury, Maryland, I had confidence in my ground track. My range along the track depended on my navigation skills and the accuracy of my wind information. Unfortunately, Salisbury had no tactical air navigation (TACAN) system at this time.

The night of the operation, my photographer appeared with his high-speed camera and support equipment. We selected a location over the water, taking into consideration the best camera angle and the challenge of taking photos from the back seat of the T-33. As we strapped in, the photographer grumbled about the color film that they had given him for his high-speed camera and the difficulty of a night operation. The film had a slower speed than he thought necessary, but the managers refused to special-order the more expensive high-speed film for the mission. Perhaps they lacked confidence that we would see the reentry at all.

We took off from Langley and climbed toward the target area. I checked in with the ground controllers at Wallops and entered a holding pattern at Salisbury VOR. Washington Center coordinated the mission. The test range was declared "hot" so that no other air traffic could enter the area. The ground-controlled countdown stood at 12 minutes and holding. I turned in toward my outbound heading and picked up the countdown as I passed over the VOR. The missile launched right on schedule as we were 12 minutes out to sea. I had the test clock set on my control panel. As we approached 30 seconds to reentry time, I told the photographer to start filming in the pre-determined direction. The timing was critical in that the high-speed camera, even with a large spool of film, only had about three minutes of film time.

I couldn't believe the results. Within five seconds of the planned time, a small light appeared in the sky similar to a star. It quickly grew brighter, changing from a dim white light to a bright green and blue glow. Just before it burned out, we could actually see the shape of the last stage. The rocket had a bolt on one end with colorful ionization flowing from it. It burned out and vanished almost as quickly as it had appeared, probably lasting only about 10 to 15 seconds at the most. I sat there for an instant, sort of spellbound. It had all worked out so well. I had a feeling of exhilaration that we had been part of it. I asked the photographer if he got his pictures, and he answered in the affirmative.

We informed the ground control station of our success and headed for home. We felt on top of the world until about noon the next day when the film was developed. Nothing showed up. The low-speed film had not been sufficient for the job. The photographer and I were interviewed and we drew sketches and described colors, but we did not have it on film. The managers decided that they could afford the high-speed color film. We tried on three other occasions to get the pictures, but never succeeded. The missile had a wide reentry footprint and we had apparently been very lucky on the first try. We learned a valuable lesson. Any operation is only as strong as its weakest link. After investing such time and effort (not to mention money!) on a high-priority mission, it is important not to skimp on any item, however small.

Another T-33 operation supported the "Little Joe" project in late 1959 and early 1960. Designed by Max Faget and Paul Purser, Little Joe was a 50-foot-tall rocket propelled by four large (for that day) solid-fuel boosters to altitudes in excess of 50 miles. It carried a boilerplate Mercury capsule to test the recovery parachute system. A few missions carried instrumented capsules with live Rhesus monkeys as passengers. A commercial manufacturer developed the parachutes under contract and NASA conducted the testing at Wallops Island.

NASA photo MSFC-9141924

The Little Joe rocket, launched from Wallops Island, carried a boilerplate Mercury capsule to test the recovery parachute system. Mallick piloted a Lockheed T-33A with a photographer to document deployment of the parachutes.

My job with the T-33 photo plane involved rendezvous with the capsule at 35,000 feet. The photographer had to take pictures of the parachutes as they deployed. There were ground cameras involved too, but the engineers wanted close-up pictures as well. The Little Joe launch was a little too complicated to give the T-33 pilot control of the countdown. We orbited the T-33 just south of the launch pad at 35,000 feet and waited. At liftoff, we turned to the outbound heading and held course for a number of seconds and then dropped the left wing into a 90-degree bank and watched Little Joe burn its way rapidly to our altitude. We stayed in the bank only a few seconds in order not to cross the rocket's path. It was a tricky operation. We had about two or three quick looks as Little Joe passed prior to parachute deployment. It was sort of like having a surface-to-air missile coming after us.

Several pilots shared this mission, and that was fine by me. I never liked the idea of having Little Joe climbing up under my belly during those long seconds that I had to maintain my course. Unfortunately, like the five-stage rocket, there wasn't much sophisticated guidance involved. It was just another skyrocket trying to go where it was aimed. On my second Little Joe mission, I rolled to the 90-degree position to confirm the launch and then back on course. My next rapid roll revealed that Little Joe had "changed its mind" about coming to my altitude and was flying horizontally at about 15,000 feet, heading out to sea. The parachutes deployed at about 150-percent of the design loads, but sustained little damage. The malfunction resulted from a solid booster failing immediately after liftoff. This unsuccessful launch actually provided enough data to bring an early end to the program. Its demise brought no tears to my eyes.

The workload in the pilot's office grew quite a bit with the normal flight test programs, logistic support flying, and special projects such as the T-33 photo missions. Two NASA pilots from Cleveland, Joe Algranti and Harold "Bud" Ream, transferred to Langley to help with the effort. Both were good aviators and fit in quite well with the operation. In fact, both later transferred to the space program when it moved to Houston, Texas. Both prospered there. Joe later served as head of flight operations at Johnson Space Center for many years.

I worked on several flight research programs during this period. I enjoyed all my flying, but research flying was the most challenging and rewarding to me. The rest was "filler-type" work, something to do while waiting for the research aircraft to be readied for the next flight.

I was soon assigned to the JF-86D. The aircraft had not flown at all for two years, while it underwent extensive modifications. A series of electric computer-controlled, motor-driven, shaker devices were installed throughout the span of both wings. The shakers provided "excitation" to the wings in order to study the aircraft's natural frequencies and responses. We established test conditions at Mach 0.9 at 30,000 feet. A complicated control panel for the shakers was mounted just forward of the control stick. There also were controls to set frequency sweeps and amplitudes. The JF-86D did not have much internal fuel capacity and the wing tanks had been removed to reduce the drag and allow us to achieve the required Mach number.

The engine had an electronic fuel control. It required a lot of time and fuel to check out this system prior to flight, so I established a procedure. I parked right by the fuel pit, cranked up the engine, and went through all of the checks on the electronic engine control. I then shut the aircraft down, keeping external electrical power on the aircraft and had it refueled. Having completed this second refueling, I called the tower for a priority takeoff. When the tower confirmed that I was cleared for takeoff, I immediately cranked the engine, taxied at high-speed to the runway and took off in an afterburner climb. As I reached the test conditions, I brought the complicated shaker system on line and performed my test points. I normally would not pull the engine out of afterburner until the last minute (i.e., about 700 pounds of fuel remaining) while heading back toward Langley. I then called for a minimum-fuel landing and made an idle-power, straight-in, approach to landing. The control tower people were familiar with the aircraft and its limitations. I talked with them by phone daily during the project.

I completed the flight test program in four months. By the time I was nearly finished, I had become proficient in adjusting the shaker system. I told the program engineer that it was like flying a jet airplane and playing a piano at the same time. It seemed as if I had about 88 different pots and switches to handle while operating the shakers. Additionally, on the resonant frequencies the aircraft and instrument panel shook so violently that I could not read the instruments. Each test flight lasted about 25 minutes. Following most flights, I shut the engine down with about 300 pounds of fuel remaining. That is not much fuel for jet operations but it was necessary in order to get enough data on the flights. The test program that took almost two years to prepare was flown and completed in five months. Operating the aircraft and shaker system and collecting the data was a challenging and complicated task, but immensely satisfying.

Along with all this fun and flying, I also had a great home life in Virginia. Audrey and I rented a small home during our first year. In 1958, we purchased a ranch home located north of Langley in the small town of Tabb, Virginia. In May 1959, Audrey gave birth to our second child, a son, Donald Karl. We rejoiced in our new home, new baby, new friends, and the country lifestyle of southern Virginia. Audrey seemed to enjoy being a homemaker. Our daughter, Sandra, developed a little southern accent that we assumed she picked up from her playmates. We planted a garden in our large backyard and joined the neighbors in raising fresh vegetables. The end of our backyard abutted deep pinewoods that extended for miles. During the hunting season, I was known to disappear after work into the woods and hunt until dark. We were well on our way to becoming Virginians.

5

Super Crusader

New Opportunities

Prior to reporting in at NACA Langley in 1957, I left a Navy fighter reserve squadron in Jacksonville where I was flying the F9F-6 Cougar. I had every intention to continue flying in the reserves, but when I arrived at Langley, I ran into a problem. Three pilots out of the current five in the Langley pilot's office were members of the same Navy fighter reserve squadron at Norfolk. Bill Alford was the skipper of the squadron and Robert Sommer and Bob Champine were both members of that squadron. Each year, when the summer drill took place, Jack Reeder was short three out of five pilots for two weeks! Had I joined, it would have been four out of six. Jack asked me not to enter the fighter squadron, but perhaps to join some other squadron in the reserve. Since NACA flying was my Number One priority, this request did not seem out of line.

In July 1957, Bill Alford arranged for me to enter an anti-submarine squadron that was flying the Grumman S2F, a small, twin-engine aircraft designed to track down and destroy enemy submarines. After flying the Banshee and the Cougar, I wasn't too excited about flying the S2F. Still, the new mission seemed challenging and I intended to approach it with enthusiasm.

The anti-submarine warfare (ASW) squadron was heavy with senior officers, mostly pilots who came from fighter squadrons. Before a pilot could check out as airplane commander in the S2F, he had to have 200 flight hours in the aircraft. This seemed a little long to me. For years, I had checked out as plane commander more quickly in other types. I think the "old heads" made that rule to keep the younger pilots out of the left seat for as long as they could. In reserve flying, it could take several years to accrue 200 hours flying time. It bothered me that I might have to wait a long time for my chance at the left seat. Setting my concerns aside for the moment, I concentrated on learning the anti-submarine mission.

The Navy had several S2F ground simulators and various ground school classes for training. One of the aircraft's primary tools for locating and tracking submarines while they were underwater was a sonobuoy system, a floating listening device that was dropped from aircraft into the water. The sonobuoy carried an underwater microphone to listen for the sounds of a submarine's propellers and transmit them back to the aircraft. By dropping a pattern (usually three) of these sonobuoys, it was possible to hear a submarine and plot its course. In this way, the S2F pilot could set up a depth-charge run to bomb the sub.

The simulators available to the Navy Reserve at that time were not terribly sophisticated, so the most beneficial part of the training came later in actual flying missions. After a reasonable amount of ground training, I was assigned my first sonobuoy mission. I flew in the right seat during training. My instructor was a senior pilot, a lieutenant commander in the squadron's operations and training department. As we flew over Chesapeake Bay, we spotted a surface ship. Since no submarine was available, we used the surface ship as a training target and dropped a pattern of sonobuoys to listen for the sounds of the ship's propellers as it made its way past or through our sonobuoy array.

The instructor pilot flew the drop pattern and released sonobuoys at the appropriate positions in the water. As we flew an orbit in the area and watched the ship proceed toward our array, the instructor switched through the different listening stations on the radio selector. He noted that all was well and that the sounds we heard indicated exactly what we were seeing with the surface target. He then asked me to interpret the scene below by listening to the sounds from the sonobuoys. I couldn't make out a damn thing from listening. There was just a bunch of "hash" and static on all of the channels. I couldn't hear any propeller noises, let alone interpret which were getting stronger and which were getting weaker, to determine the track of the target. After a few minutes, the instructor pilot seemed to grow a little exasperated that I was not able to make anything out of it. Again he switched through the various channels and verified that we had a good sonobuoy response. I felt badly about the mission because I hadn't gotten a thing out of it and I believed that it was due to my own ineptness.

After we landed, taxied to the chocks and shut down, I walked to the rear of the aircraft for a general post flight inspection and discovered that our three sonobuoys were still very much in place. The launch system had failed and we had not dropped one in the water. The instructor pilot had made up the whole damn scenario either through his imagination or desire to impress me!

I felt better knowing that I had not succumbed to his pressure to visualize the scene below from sounds that did not exist, but I was extremely upset about the fact that this senior instructor pilot from the squadron could have made such a mistake. I considered approaching the squadron commander about it, especially because the instructor pilot did not seem too disturbed by his obvious screw up. In the end, I decided not to say anything. A month or so later, I reviewed my workload at Langley and my disillusionment with ASW flying. I dropped out of the squadron. I couldn't picture myself flying 200 hours in the right seat for this particular operation.

At first, Bill Alford was quite upset when he learned that I had quit the squadron. Once I explained the situation to him, he didn't argue about it. He did insist, as a friend and senior officer, that I remain in the Naval Reserve.

There were several Navy Reserve transport (VR) squadrons at Norfolk that flew R4D and R5D, the military equivalents of the civilian DC-3 and DC-4 transport aircraft. My flight experience in the C-47 and T-29 qualified me to enter the VR squadron, but at the time they were full of reserve pilots who also worked for the airlines. There was no room for me. At Bill's urging, I joined an engineering unit, called a Bureau of Aeronautics Reserve Training Unit (BARTU), also located in Norfolk. I remained in the BARTU until an opening came up in the transport squadron in May 1960. Most of my flying in the VR squadron took place at Norfolk. I gained a good deal of experience there, flying with a number of seasoned airline pilots, which actually helped me my support flying assignment at Langley.

Shortly after I joined the VR squadron, the Navy shut down the fighter reserve operation in Norfolk. Jet noise had become a problem as neighborhoods grew around the base. The Navy Reserve fighter pilots in the NASA Langley pilot's office were transferred into the VR squadrons. This worked out well because there were several VR squadrons. During the summer cruise deployments, only two pilots were now absent from Langley flight operations at one time.

Flying Qualities of the F8U-3

In June 1959 I was assigned to a new high-performance aircraft program at NASA Langley. The F8U-3, called the "Super Crusader" or "Big Crusader," was much larger than the F8U-1. Powered by a Pratt & Whitney J75 engine that provided over 26,000 pounds of thrust, it could attain speeds in the Mach 2 range. It had a similar general configuration to the earlier F8U-1 in that it had a variable-incidence wing, but the engine inlet was much different. There was a scoop on the lower lip of the inlet to properly align the shock wave at high speed. There were also manual-control bypass doors to bleed excess air overboard so as to prevent inlet choking at supersonic speeds. A long, sharp nose cone and two large folding ventral fins made the F8U-3 distinctively different from its predecessors.

Vought Aircraft built the F8U-3 in competition with the McDonnell F4H Phantom II. The Navy selected the F4H over the F8U-3 as its next fleet fighter. Both aircraft were flown in an extensive evaluation program prior to selection, but the F4H won the "fly-off" because of its multi-engine, two-place configuration. There had been times when a multi-engine fighter landed safely on one engine while single engine aircraft were often lost due to engine failure. The Phantom II's versatile design made it the choice of all the services. It provided decades of service to the Navy, Marines, and Air Force. The F8U-3 had better speed and turning performance and had less drag, but the F4H had greater ruggedness and survivability and was better suited to carrier operations.

After the fly-off competition, two Super Crusaders were made available to NASA for research purposes. These airplanes had speed and altitude

The Smell of Kerosene

NASA photo L-59-6101

The Vought Aircraft F8U-3 was capable of attaining altitudes above 50,000 feet and speeds exceeding Mach 2. Mallick flew the aircraft for sonic boom studies at Langley.

performance that made them ideal for investigating the sonic booms that were created by all aircraft flying faster than the speed of sound. The shock waves that followed along behind supersonic aircraft sometimes broke windows and caused structural damage to buildings.

Bill Alford and I shared flying duty in the F8U-3 program. We were both pleased at the opportunity as this would be the highest performance aircraft ever operated at NASA Langley. Because the aircraft were prototypes, there were no operational simulators or ground schools, or even system mock-ups. Training was provided by Vought chief test pilot John Conrad, who ferried two of the F8U-3 prototypes to Langley. The first arrived on 26 May 1959. It served as the primary flight-test aircraft. The second arrived a month later and was mostly used for spare parts. John conducted a one-week ground school to teach me and Bill about the aircraft's systems and handling qualities. John provided flight manuals and personal notes regarding lessons learned from flight test experience with the aircraft. I found the entire program to be an exciting challenge.

The aircraft was easily capable of exceeding 50,000 feet altitude and we planned to conduct tests above 60,000 feet. Therefore, Bill and I arranged to visit the Navy's Full Pressure Suit Unit at Norfolk, Virginia. There we were fitted and checked out in the latest high-altitude life support gear. The suits were rather crude by today's standards, but they were adequate for the job. At

Langley we did not have personnel familiar with pressure suits and associated support equipment. We assigned several of our ground crew to the task, and they joined Bill and me for training by Navy personnel. One of our suit technicians, Joe Schmidt, later joined the Space Task Group. He gained a reputation as one of their experts regarding astronaut's space suits and equipment. Almost everything about this operation was new for those of us at Langley, but our talented people soon put the program in good order.

This was my first experience flying in a full-pressure suit and it was memorable. The suit was made of several layers, the inner one was a rubberized, air-sealed material that provided a reservoir of air pressure when flying at extremely high altitudes. Next was a strong nylon layer that kept the inner rubber suit from expanding into a round ball when pressurized. When the suit was inflated to an operating pressure of around 3.5 pounds per square inch, it was like being inside a football. The suit expanded as far as the outer garment would allow and the pilot felt like a large bear, barely able to bend his arms or legs. In later suits, mobility was improved. The space program significantly advanced pressure suit technology. Unlike today's space suits, our Navy suits did not have any self contained-cooling. We relied on the aircraft's air-conditioning system, plugged into the suit, to provide cooling. We also had an aircraft-to-suit oxygen supply. Our early Navy suits had a small emergency oxygen tank located in a backpack that provided oxygen and suit pressurization in the event that we had to eject from the aircraft. If a pilot had any tendency toward claustrophobia, the full-pressure suit would bring that out. When I first started to wear the suit, I couldn't believe how restrictive and cumbersome it was. I wondered how tough it would be in air-to-air combat maneuvering as the pilot's ability to look behind him was limited compared to flying in a regular cotton or Nomex coverall. Like most uncomfortable things, one gets used to them in time. I gradually became accustomed to wearing the suit.

Our checkout pilot stayed with us until Bill and I had flown two flights each. John was on the radio during our checkout flights and he provided answers to our questions on a real-time basis. It helped a great deal. After each flight, John debriefed us. Bill and I described the flight and any problems we observed. We asked questions and John shared his knowledge of the aircraft. The checkout flights went very well. Soon, John headed back to Texas and Bill and I began the flight research program.

During tests, we flew a track that started east of Virginia Beach and crossed directly overhead of Wallops Island, just a little southeast of Salisbury, Maryland. The test range was about 100 miles long. NASA technicians established a line of sensitive microphones along the final miles of the track, and they had a perpendicular line of microphones across the flight path at Wallops Station. Many of the precisely positioned microphones were installed on boats on the ocean. These recording stations extended about five miles east and west of Wallops Island.

The Smell of Kerosene

My mission plans required me to fly at a pre-determined speed and altitude as I entered the test area. I had to be stabilized on the test conditions at least a minute prior to entering the so called "gate" in order to assure accurate data on the ground. We established a baseline speed of Mach 2.0 for all altitudes from 35,000 feet to 60,000 feet. The F8U-3 was easily capable of reaching Mach 2.2, and had even reached Mach 2.38 on a speed run at Edwards Air Force Base in California some months earlier.

The F8U-3 had better performance than any aircraft flown to date. It was a big fighter, almost 40,000 pounds with full fuel, and had a tremendous engine. I was extremely impressed during my first takeoff. I accelerated to "military power," released the brakes and lit the afterburner. The aircraft lurched forward, rapidly accelerated, and I passed through rotation speed almost before I realized it. After takeoff, I had to pull the nose up quickly to avoid exceeding safe limits for gear-down with the wing in takeoff position. I pulled back on the throttle, taking the engine out of afterburner, once I was safely airborne and retracted the gear.

Unlike earlier model Crusader's, the F8U-3 had an extra pair of airfoil surfaces. There were two folding ventral fins located on the lower fuselage near the tail of the aircraft. When I lowered the landing gear, the ventrals assumed a horizontal position similar to an extra horizontal tail. This prevented the fins from striking the ground during takeoff and landing. Once the aircraft was airborne and the landing gear retracted, the fins moved to a vertical position on each side of the lower fuselage. They provided additional directional stability at higher speeds. Wind-tunnel and flight test data had demonstrated the problem of reduced directional stability of a vertical tail surface as Mach number increased. The ventrals added more drag, but the large powerplant overcame this problem with ease.

On my first takeoff, I rotated the nose and was immediately airborne. As per the briefing, I quickly moved the landing gear handle up and rotated the nose even higher in pitch attitude to keep from exceeding the maximum gear-extended speed. As the gear went up and the movable rear ventrals went down, the aircraft wallowed a little. I had discussed this anomalous motion with John Conrad prior to flight and it came as no surprise. I was surprised, however, that the aircraft had a distinct sideslip. A post-flight discussion with John later revealed that the afterburner nozzle on this aircraft actually distorted from a circle to a slight oval when the afterburner was lit, and the thrust vector was slightly misaligned with the aircraft's center of gravity. This undesirable characteristic only manifested at lower speeds, such as takeoff with the afterburner engaged. At higher speeds, the effect was scarcely noticeable.

After takeoff, I climbed in military power to about 25,000 feet. I lit the afterburner climbed to an altitude of around 30,000 feet. The F8U-3 had excellent handling qualities. It had light and comfortable control forces, as a fighter should, and there was no tendency to over-control or cause pilot-induced oscillation (PIO) in any of the three axes of control. During the climb, I maneuvered to a point just off the coast of Virginia Beach and checked in with the Ground Control Intercept (GCI) facility that served as air-traffic control. I also called the Wallops Island radar controller to check on the status of the test range. Once the Wallops controller gave me the go-ahead, I turned north and began my acceleration to altitude and speed. I can remember the response of one GCI controller as he followed my track north, and I quickly accelerated to Mach 2.0. "Sir," he asked, "what in the hell are you flying?" He was unaccustomed to seeing such performance on his radar screen, and my rapid acceleration and high speed immediately caught his attention.

The inlet controls were manually controlled and adjusted with regard to Mach number. The bypass doors opened at about Mach 1.4, accompanied by an increase in thrust. They bled excess air overboard to allow the engine and inlet to operate more efficiently by converting the inlet air's velocity to pressure. A problem with this prototype aircraft's propulsion system was that the pilot could not adjust the thrust during afterburner operation. It was either in full afterburner or no afterburner, with no setting in between. Had the aircraft been selected for production, I'm sure that characteristic would have been corrected. For our operation, however, we had to live with it. To keep the aircraft from accelerating beyond the desired speed of Mach 2.0, I deployed the speed brake to hold my velocity. It always struck me as counter-intuitive that I controlled the aircraft's speed into the test "gate" by using more or less brake extension. I had never flown an aircraft with excess thrust. The use of speed brake for speed control added another task for the pilot.

The F8U-3 also had a problem with compressor stalls. On a high-speed supersonic aircraft, the relationship between engine and inlet is important. A problem that affects one will immediately affect the other. On the F8U-3, engine stall occurred without warning during a supersonic run. I had no method to restart the inlet or clear the stall until the aircraft had slowed to subsonic speeds. Both Bill and I experienced stalls, and they were startling and dramatic. I can recall one stall I experienced as I accelerated through Mach 1.6. The engine inlet was located right under the floor of the cockpit, and the stall hit with a loud thump. My feet lifted right off the cockpit floor! This was just the beginning. The inlet started to buzz, shaking and rattling the aircraft as it slowed down. I moved the throttle out of afterburner to the military thrust setting and waited. The F8U-3 had to slow to Mach 0.9 before the engine stall cleared. Fortunately, engine temperature did not seem to be a problem and I did not have to shut the engine down. Had the engine exhaust gas temperature started to climb, my next move would have been to shut the engine down and do a little gliding. Luckily I didn't have to do that. The F8U-3 wouldn't have made a very good glider! After the stall cleared itself, I flew another pattern and determined that the aircraft was sound. I then made a successful supersonic run, but only one this day. We normally made two Mach 2 test runs on a full load of fuel.

Courtesy of the Air Force Flight Test Center History Office

The F8U-3 Super Crusader (right) was distinctly larger than the F8U-1 Crusader. Its modified nose and inlet are also apparent.

During the program, we also experienced some difficulty with personal equipment and support. We normally flew two missions each day. The same pilot flew both missions because of the difficulty associated with getting in and out of the full-pressure suits. It just seemed easier for one pilot to fly both missions and keep his pressure suit on. After the first flight, I usually taxied to the fuel pit. I stayed in the aircraft while it was refueled. A sun shield was rolled up to the aircraft to shade the cockpit. The ground crew supplied me with lemonade or water. The pressure suit had almost no cooling capability during ground refueling because we did not have portable liquid cooled oxygen tanks available to hook up to the pilot. We only had a pressure reducer unit that we built to take compressed air and flow it through the pilot's pressure suit during the refueling. This helped somewhat, but flying these missions in the middle of the summer in Virginia was very uncomfortable. After refueling and engine start, the plane's air-conditioning system provided adequate cool air. In fact, sometimes it was a shock as cold air poured in over my sweaty body that had just been broiling in the heat. After completing my first or second day of flying, I decided to weigh-in before the mission and weigh-out afterward. I found that I lost five pounds over the course of two flights. When I removed the pressure suit perspiration ran out onto the floor and made quite a puddle. Fortunately, by drinking fluids in between flights and using the compressed air, I did not suffer too much dehydration.

Super Crusader

For several reasons, we did not use a chase plane on the sonic boom missions. We had no other aircraft that matched the speed and altitude of the F8U-3. Also, the scientists measuring the sonic boom overpressures did not want the interference of a second aircraft that would alter the data. Lastly, even though the range was 100 miles long, it was just off the coast and always within radio range of our control center. Therefore, Bill Alford and I supported each other on the missions. If Bill was flying, I was on the control radio, and he served as controller for my missions.

The pilot's windscreen, the portion of the canopy directly in front of him, was unusual. Almost all fighter aircraft, at the time, had a piece of thick bulletproof glass installed directly in front of the cockpit. The F8U-3 did not have this. It had only standard Plexiglass like the rest of the canopy. This limited our cruise time at Mach 2. High-speed airflow easily heated the thin plastic material, structurally weakening it. After 10 minutes, its reduced strength left it vulnerable to catastrophic failure. We therefore set a five-minute limit on cruise at Mach 2.0. This gave us a 100 percent safety factor. The test conductor kept an eye on the time spent at cruise conditions. Once the pilot reported reaching Mach 2.0, a clock started on the ground and the pilot was given a call when he had been at that speed for five minutes. The ground controller gave the pilot precise information regarding ground track, altitude, and airspeed. The pilot was busy controlling this "thundering monster" into the gate on-speed, on-altitude, and on-track. One of the problems that we had early in the program was training our own radar people to give the pilot useable instruction on track corrections. Initially, we found them calling for 30-degree heading corrections one mile from the gate at Mach 2.0. This was impossible for the pilot to accomplish. After just a few flights and extensive ground debriefings, we worked as a team and the data began to accumulate.

The five-minute limit at Mach 2.0 was not a problem except on a "cold day" at altitude. Air temperature affects aircraft performance. A hot outside air temperature results in much slower acceleration and longer takeoff distances. The reason for this is that the jet engine develops thrust by accelerating a mass of air from the front of the engine out the back. The cooler the air, the heavier it is and the engine, with its set volume of airflow, is able to develop much higher thrust on a cold day due to the increased density of the cooler air. The same is true for acceleration at altitude. If the outside air temperatures are colder, the engine is able to develop more thrust and accelerate to a given speed in a much shorter time and distance. On several occasions, we experienced colder-than-normal days at altitude and we reached Mach 2.0 very quickly. Getting to the test gate prior to exceeding our five-minute cruise limit was a struggle.

Bill and I became quite a team. We not only worked well together, but we were good friends away from work too. Bill was an avid bird hunter, and with my background from the Pennsylvania woods, it was natural for us to become

The Smell of Kerosene

hunting buddies, too. Bill had also been my check pilot on a number of airplanes, and my success reflected on his skill as an instructor. Bill and his wife had a nice country-style home with a large back yard. They hosted a number of get-togethers and barbecues for people from the office and their families as Bill enjoyed having the group together.

Bill and I exchanged every piece of information that we thought was important on the F8U-3. If something surprised me about the airplane or caught me short, I passed it on to Bill and he did likewise. We had several incidents during the operation of this relatively new and high-performance airplane that grabbed our attention. On one of Bill's flights, an air-pressurization fuel-transfer hose disconnected from a 4,000-pound fuel tank in the wing that provided a big percentage of our fuel. Bill was busy with the acceleration, preparing to run the gate, when he noticed the fuselage feed tank was desperately low. He immediately pulled the engine out of afterburner and started a rapid turn back toward Langley Field. As the big jet passed over residential areas, a sonic boom rattled windows and nerves in Salisbury, Maryland. Fortunately, no damage was done and Bill made an emergency "low fuel" landing at Langley with about 350 pounds of useable fuel left and 4,000 pounds of unusable fuel in the wing. That was the closest we came to losing the aircraft and I was glad that I didn't have to handle that one.

On one of my flights, I forgot to turn up the heat to defog the canopy as I started my descent for landing. I normally had the heat turned low and the air-conditioning on cool during accelerations and high-speed runs. Once I started my descent, however, I turned the defog heat up to full to prevent the warm moist air at low altitude from condensing and fogging the canopy as I descended. It may not seem like a big deal, but it was. The air was so moist in the summer that it took some time, flying at lower altitude with the heat turned full-on, to clear the moisture off the inside canopy so I could land. There was so much moisture in the cockpit that it looked as if a summer rain shower had passed through. I used a handkerchief and my pressure suit glove to wipe a clear spot on the windscreen during final approach. Since I usually made two Mach 2.0 runs per flight, I had little fuel left for pattern operations. I briefed Bill about this moisture problem so that he wouldn't learn about it the hard way.

On another occasion, I made an afterburner takeoff and climb to altitude. This used up a lot of fuel and it was uncomfortable because the nose attitude was so high in the initial climb that pilot visibility outside the cockpit was extremely limited. I wanted to make the afterburner climb, however, for my own experience. Because I did not keep the nose attitude high enough, I actually went supersonic over the residential area of the peninsula. Fortunately, no damage was done and there were just a few jarred nerves from the sonic boom. It was hard not to be impressed with the performance of this aircraft.

The airplane's flying characteristics were excellent in almost all phases of flight. In the landing approach configuration, the aircraft was extremely speed-

Courtesy of the Air Force Flight Test Center History Office

In flight, a pair of ventral fins unfolded to give the F8U-3 additional directional stability at higher speeds. The fins folded up prior to landing.

stable. When I changed the nose attitude, the altitude also changed, but the speed tended to remain where it had been before the attitude change. I think that this was due to high drag caused by the wing in the approach configuration. The F-8 aircraft were built with a two-position wing to provide increased visibility at low takeoff and landing speeds. The system allowed the pilot to increase the wing angle-of-attack without increasing the fuselage angle. It made for a very comfortable controlled approach. As the wing tilted upward, however, it also exposed a flat-plate structure on the front of the wing assembly to the airstream. This added noticeable drag.

The F8U-3 had a few drawbacks, including one that involved its crosswind landing characteristics. It had a tremendous fuselage side area and a very narrow landing gear footprint. The aircraft leaned away from the wind during landing, even when I applied a large amount of aileron deflection. It made me feel very uncomfortable until the aircraft slowed to low speed. Another problem with the aircraft was lack of an anti-skid system on the brakes. If I used too much braking while over a speed of 80 knots, there was a good chance of blowing a tire by locking a wheel. On a crosswind landing I just had to ride out the lean, keep it straight on the runway and wait until I slowed below 80 knots before starting to apply light breaking. If I lowered the wing back into the fuselage after landing, it reduced not only the lean but also the drag. This made the aircraft even harder to slow. Bill and I decided the wing should stay up to give us the drag we needed, even though it contributed to the lean of the aircraft in the crosswind. These problems were non-existent for a no-wind or headwind runway condition. The aircraft handled very nicely then.

By early October 1959, Bill and I had made good progress with the flight tests. At that time, Bill had an opportunity to travel to England to evaluate a new airplane. The Blackburn NA.39 Buccaneer was a two-place, carrier-borne strike aircraft that had excellent range and was capable of delivering a nuclear weapon. Powered by twin turbojets, it featured an area-ruled fuselage and a boundary-layer control system for its wing and tail surfaces. As one of our senior research pilots, Bill had been selected to evaluate the aircraft, which had not yet entered operational service. I was pleased for him and we discussed my plans to continue work with the F8U-3 in his absence. The F8U-3 test team was up to speed and there was no reason not to forge ahead. Bill's opportunity meant more F8U-3 flights for me, so it was a bonus to both of us.

Bill had been in England for about a week and I was charging ahead on the F8U-3 program. I had just completed a successful test flight, changed out of my pressure suit, and showered. As I took the stairs two at a time up to our second-floor pilot's office, I was on the "high" that I normally experienced after flying the F8U-3, especially when the mission had gone so well.

As I burst into the office, Jim Whitten and Bob Sommer were sitting at their desks, and I proceeded to expound on the aircraft and how impressed I was every time I flew it. I had never experienced such performance and handling qualities. I noted that Jim and Bob were not too attentive, and once I stopped talking, I found out why. Jim looked at me and said "Don, Bill Alford was killed this morning in a crash of the Blackburn 39."

My world suddenly came to a halt. I couldn't believe it and yet I knew that it must be true. I have never gone from such a "high" to such a "low" so rapidly. Suddenly, my accomplishments in the F8U-3 didn't seem important anymore. I had been working very hard on the tests. In the back of my mind, I was motivated to make Bill proud of me and satisfied that he had prepared me well for the task. Now he was gone. Just like that. Gone.

I made my way to my desk and all three of us sat for several minutes, not speaking. We each pondered our separate thoughts and memories of our fallen comrade. I asked if there were any details about the accident, not that it would change anything or bring Bill back. Bill and a British flight test engineer were evaluating the Buccaneer's low-speed flying qualities when it departed controlled flight and crashed. Jim said that poor weather had forced them to operate at a lower altitude than desired and they were too low to eject safely from the aircraft. Both men perished.

I had been exposed to aircraft crashes and pilot's deaths from the very beginning of my Navy flying career. I had known a number of people who were killed flying. I had built up a "shell" for myself over the years. I accepted this as part of the profession, but it always happened to someone else. It was always "the other guy." My protective shell was broken by Bill's death. The Langley pilot's office was a small, tight-knit group. I was the sixth pilot to join the group and Bill was the one that I most closely related to. I was very shaken and it took me a long time to recover.

There have been accidents from the very first days of flight and there will be as long as man flies. Flying an aircraft is a complex task technically, physically, and mentally. Naturally, aircraft operators always strive to limit the number of accidents. The combination of unique flying machinery and the man-in-the-loop (pilot) makes it virtually impossible to eliminate accidents altogether. Some result from mechanical failures in which a critical part breaks down. Others result from pilot error. Usually, it is a complex combination of factors or a chain of events. If any one element in the chain were different, the accident might have been avoided. Unfortunately, such knowledge is usually gained only in hindsight.

I was profoundly shocked by Bill's death for a number of reasons. First, he had become a very close personal friend of mine. I liked his personality and the way he looked at life. I admired the way he strove for excellence, always exceeding normal expectations. He encouraged me to strive for the best in my own career. I respected Bill and his talents as a research pilot. When I flew with him on checkouts of various airplanes and helicopters, I marveled at his proficiency. I strove to emulate him.

I reviewed the accident details as they became available and tried to determine what had induced Bill to conduct the low-speed evaluation tests at such a low altitude. Knowing Bill, I can only surmise that his dedication and sense of duty were his undoing. Despite the adverse weather, he must have decided to complete the evaluation as planned and return with the data. I told myself that his desire to perform the task on schedule had inadvertently compromised his safety. I needed to find an explanation as to why this accident had occurred. I had to do it to rebuild my protective shell.

Bill's body was flown back to the United States soon after the accident. He was buried in Arlington National Cemetery in Washington, D.C. with full military honors.

Langley Flight Research Gains New Focus

By the end of 1959, the office workload was high. This helped to put the Bill's memory toward the back of our minds, but never out of mind completely. Langley's VSTOL-type research programs increased during this period, bringing new challenges. Jack Reeder, Jim Whitten and Bob Champine did the majority of the advanced helicopter and VSTOL flying. Jack was outstanding in this area of flight test and received national recognition.

As the NASA organization grew, new blood came in from other agencies and the military services. Less that two years after the formation of the agency, it became quite evident that space flight had the top priority. A gentleman named Richard E. Horner joined NASA headquarters as associate administrator in June 1959. After a short study of the NASA organization, Horner decided that the various NASA centers were duplicating too many of their flight research efforts. He directed that henceforth Langley would serve as primary center for VSTOL and helicopter testing. High-performance flight research, as we had performed with the F8U-3, F-101, F11F and others, would now take place at the Flight Research Center in California. Like many Langley aeronautics people, I did not agree with Mr. Horner's analysis. Although a number of NASA centers worked with high-performance aircraft, specific programs were not duplicated. There was an overall aeronautics effort spread across a number of centers, just as there had been for years in the old NACA. This had never been criticized before, but now that space research priorities dominated the agency, aeronautics research efforts were being trimmed. Horner, having made his mark, stayed with NASA for just over a year. He left the agency in July 1960 to take a job as Senior Vice President of Northrop Corporation.

The reorganization hit me hard because I enjoyed high-speed flight research. As much as I liked the other flying, such as helicopters and support, nothing gave me greater satisfaction than flying a high-performance program. There is no question in my mind that if NACA had not been converted to NASA, this particular change would not have occurred. In any case, there were no options at this time and I accepted my fate. I would make the best of a bad situation. It was obvious to me now that the NACA was truly gone, absorbed into NASA. It may have originally been the nucleus, but its identity and control had disappeared within a year and a half. It seemed to me that they should have named the new organization the National Space and Aeronautics Administration. It would have made the point that aeronautics was now playing "second fiddle" to space.

At any rate, I soldiered on. In February 1960 I was assigned to an interesting engineering program. For a number of years, the NACA had instrumented a number of commercial airliners with small VGH recorders. They logged the aircraft's velocity (v), gust and maneuver acceleration (measured in "g-units"), and altitude (h), hence the acronym. The study provided data to predict the accelerations and loads that an aircraft might experience in its lifetime. The VGH recorders were installed on a few aircraft with most of the major carriers. The final data provided aircraft designers with information that helped determine aircraft service life and load limits. Metal fatigue was of primary concern. Over the years, the NACA and then NASA experienced a fine relationship with the various airlines involved. The Federal Aviation Administration (FAA) was aware of the program, but was not directly involved. The NACA, and later NASA, was pleased with this arrangement in that the

Boeing photo A-70332

Shortly after the Boeing 707 entered service, NASA engineers began to collect performance data. Mallick and engineer Tom Coleman used the data to brief commercial operators on ways to improve their pilot training programs.

airlines and FAA sometimes seemed to have an adversarial relationship and cooperation was not always good. The FAA served mainly as a "watchdog" over the airlines. For years, the VGH program successfully provided data on propeller-driven airliners. Engineers gathered and documented the data and passed it on to the aircraft manufacturers.

Shortly after the Boeing 707 jets entered airline service, NASA engineers noticed an alarming new trend. The jet data showed a large number of overspeeds, instances in which the aircraft had exceeded its maximum design speed. The majority of these incidents occurred at the pushover point at the beginning of descent. It was obvious to the engineers that the pilots were flying the jets as they had flown the propeller aircraft. In the older aircraft, the procedure called for pushing the nose over and reducing power later, as required for descent. With the faster, more streamlined jets this procedure resulted in an overspeed that sometimes continued during the descent. This "out of the envelope" operation had the potential to significantly reduce the fatigue-life of the aircraft.

Clearly, the problem needed to be addressed, but it was a delicate matter. NASA planned to report that airline pilots weren't flying their aircraft within prescribed flight envelopes. Neither the FAA nor the airlines were going to be happy about this. The airlines had been very cooperative about letting us use their aircraft and crews to gather data. We didn't want to get anyone in hot water, but we had to address the problem right away. NASA selected me and a very capable engineer named Tom Coleman to write a report. Tom had been involved in the VGH program for years and was on top of the problem. First, we prepared an engineering presentation that reviewed the speed acceleration (VG) envelope of each aircraft and pointed out the significance of the aircraft's various limits. We concentrated on the normal reduction of safe acceleration levels as aircraft speed approached the high end of the envelope. We then took records from each of the subject airlines that showed cases of overspeed and coded them so that only Tom or I could identify the airline or flight number. This prevented the FAA or airlines from taking punitive action against any crewmembers. We wanted this information included in the pilots' initial and refresher training.

For five weeks, Tom and I toured the U.S., visiting the operations departments of airlines operating the new jets. Overall, I was impressed with the airline managers. The majority were senior airline pilots who had moved into management and training positions. Their response to our briefings was generally good. They agreed to integrate the NASA presentation as a guideline for their own training programs. As one might expect, there were a few airlines that absolutely denied that their pilots had ever exceeded aircraft performance limitations. In these cases, we showed them the coded data strips that identified the type of aircraft and the overspeed data. We, of course, did not identify the flight numbers involved. It was interesting to me that the ones who made the strongest denials at first were also the ones that wanted to identify the flights from our data in order to punish the pilots involved.

All in all, the briefings went well. I was glad that the airline operators benefited directly from research data they had supported. Data tapes taken over the next few months indicated a great reduction in overspeed incidents. The system worked!

Flying the Skyhawk

During the summer of 1962, the Navy decided to establish a Navy attack squadron at Norfolk, jet noise notwithstanding. The new unit was equipped with Douglas A4D Skyhawk aircraft. The Navy's main problem at this time was finding qualified jet fighter pilots in the area to man the squadron. The shutdown of the Navy fighter squadron several years earlier resulted in jet-qualified pilots going to other squadrons and flying aircraft other than jet fighters. Since Robert Sommer and I were currently flying jet fighters at NASA, the Navy asked us to set up the squadron, designated VA-861. Bob was the commanding officer (C.O.) and I was the executive officer (X.O.). We built up the squadron with younger Naval Aviators who were being released from active duty and were interested in flying in the reserve.

I accrued about 25 hours flight experience in the A4D. It was a feisty little airplane to fly. In the "up-and-away" flying, it was just as much a fighter as it was an attack aircraft, maneuverable and capable of turning on a dime. The cockpit was small, compact, and built around the pilot. All the switches and levers were right where they should be, within easy reach. Douglas Aircraft Company engineer Ed Heinemann designed the Skyhawk to be lightweight, small, fast, and capable of delivering a nuclear weapon. One of his most successful designs, the aircraft became known as "Heinemann's Hot Rod." In a way, it was like driving a little hotrod car, small but with lots more "oomph."

My evaluation of the aircraft was limited. I never had the opportunity to evaluate it as a weapons platform for gunnery, rockets, or bombs. I found no problem areas in the handling qualities. If there were any, they were probably overshadowed by the high performance.

I always looked closely at the landing approach configuration of any aircraft I flew. As an old Navy instructor told me early in my career, "Son, if you can't take off and land an airplane properly, you can't do anything else with it." I'm not sure that's completely true, as I have heard stories over the years that some of the best air-to-air aces we produced were not so smooth in the landing pattern. Perhaps they were just more interested in flying and combat tactics and thus concentrated on those aspects of the mission. In any case, I looked closely at just how nice an aircraft flew in the landing pattern and touchdown. The A4D, with its short fuselage and effective horizontal tail, was sensitive in pitch control. I found it difficult to fly a precision approach by referring only to the airspeed indicator. The Skyhawk was "clean" even in the landing approach configuration. Therefore, any change in nose position would result in a rather quick departure from the desired airspeed. Fortunately, the aircraft had a fine angle of attack (AOA) indicator, and the secret was to fly a fixed AOA and control the altitude with power adjustments. I also found the aircraft's crosswind handling characteristics on roll out difficult to handle. The Skyhawk had tall, narrow landing gear and the aircraft leaned away from the wind as the F8U-3 had, but for slightly different reasons. The first time I landed the A4D with a crosswind at the published limit, I made all the required control inputs right after touchdown. Even so, it was still a very uncomfortable rollout. As with the F8U-3, the Navy did not see a necessity for an anti-skid system, and great care had to be exercised not to blow a tire while braking under crosswind conditions. The Navy never seemed to pay much attention to an aircraft's crosswind handling qualities, figuring that most of their operational work would be from carriers where the ship turns into the wind for flight operations.

While I was having fun flying with NASA and the Navy, Audrey was very supportive. I think, at times, that she thought that I was having too much fun, but if I was happy, so was she. Audrey presented me with our third child, a son, David Glenn in October 1962. I picked Dave's middle name in honor of the Marine major who had so impressed me at Johnsville and later in the Mercury program. Audrey and I socialized with the pilots and wives from our office and we also enjoyed the Navy Officer's Club at Breezy Point in Norfolk. Because of the family growth, we were planning to buy a larger home. Although I was testing VSTOL aircraft for NASA and the Navy had seen fit to put me back into jets, I felt pretty comfortable with my life in Virginia and was satisfied with what I was doing. Then, as the expression goes, "a funny thing happened on the way to the Forum."

In December of 1962, Jack Reeder got a call from Joe Walker, the chief pilot at the NASA Flight Research Center (FRC) at Edwards AFB, California. Walker wanted to know if any of the Langley pilots would like to transfer to FRC. They needed one or two more pilots as work at the Center was picking up. When Jack walked into our office and made the announcement, I told him that I was very interested but would have to check with the "home office" (i.e. Audrey) that evening and give him my final answer the next day. Audrey and I discussed the pros and cons of such a move. I really wanted to participate in the type of flight research activities that were being conducted at Edwards, but we both loved the Virginia area and the people. We had made a lot of new friends and our kids were doing well in school. It was a tough decision. Ultimately, we decided to buy our new, larger house in California. The next morning, I told Jack that I would like to be considered for transfer. The wheels were put in motion and the transfer was arranged.

6

High Desert Flight Research – Edwards, California

Origins of the NASA Flight Research Center

In 1944, NACA Langley Memorial Aeronautical Laboratory (LMAL) became involved in the development of a new rocket research airplane called the XS-1, and later, simply X-1. The "X" stood for experimental and it was designed for flight research in the transonic and low supersonic range. Designed using the shape of a 50-caliber bullet, it was the first aircraft built for manned flight faster than the speed of sound. This was also the first time that Langley scientists and engineers had been involved in the initial design of a research airplane. NACA engineers developed the basic design criteria for a transonic research airplane, including wing airfoil, instrumentation requirements, and the horizontal stabilizer configuration. Bell Aircraft Company in Buffalo, New York, built three X-1 aircraft, each powered by a four-chamber Reaction Motors XLR11 rocket engine. Flight test operations were supported jointly by the Army Air Forces and the NACA. The vehicles were dropped from high altitude by a modified B-29 Superfortress that served as a mothership.

Bell completed construction of the first X-1, minus its engine, in late December 1945. Although the NACA hoped to perform initial glide flights at Langley, the Army selected Pinecastle Field near Orlando, Florida, instead. A young engineer named Walter C. Williams headed a small team from LMAL that gathered data from the X-1 while in Florida. Bell chief test pilot Jack Woolams made 10 glide flights at Pinecastle between January and March 1946. Next, the airplane returned to Buffalo for installation of its engine.

While the Pinecastle tests were successful, they highlighted the need for a more suitable test site for the powered flight series. Planners wanted a secluded location with good flying weather, ample landing area, and sparse population. Muroc Army Airfield in California's western Mojave Desert fit the bill perfectly. An excellent location for high-speed aircraft testing, the site included a 44-square-mile dry lakebed. It was the ultimate landing field!

The name Muroc had a unique origin. In 1910, the Corum family settled on the west shore of Rodriguez Dry Lake, adjacent to the Atchison Topeka & Santa Fe railroad Line. They initially named the site Rod, short for Rodriguez. The Postal Service rejected an attempt to rename the settlement Corum because there already was a town called Coram in northern California. The letters were then reversed to spell Muroc. Rodriguez Lake was renamed Rogers Lake.

In September 1946, LMAL sent a detachment of engineers and technicians to Muroc to support the X-1 project. Headed by Walt Williams, the NACA Muroc Unit was meant only as a temporary detachment from LMAL. In September 1947, however, LMAL established the NACA Muroc Flight Test Unit as a permanent facility. The X-1 program was only the beginning. In the quest for higher and faster manned flight, many new aircraft designs were sent to the desert for evaluation. On 14 November 1949, the NACA unit at Muroc became independent from Langley as the NACA High-Speed Flight Research Station (HSFRS). Muroc AAF underwent many changes during these years, first becoming Muroc Air Force Base in 1948 when the Defense Department established the U.S. Air Force. The following year, Muroc AFB was renamed Edwards AFB in honor of Capt. Glen Edwards who died while testing the YB-49 bomber prototype. In 1950, the base expanded its facilities with new hangars and runways. In June 1954, the HSFRS moved to a new location near the northern edge of Rogers Dry Lake and the following month was renamed the NACA High Speed Flight Station (HSFS). At this time, it had about 250 employees. NACA became NASA in 1958. In September 1959, the NASA High Speed Flight Station became the NASA Flight Research Center (FRC). This status increase, from a Station to a Center, put the facility on the same level as the other laboratories (now also designated as Centers) in the NASA organization. This boosted employee morale, as well as provided greater prestige and autonomy. Small in size, compared to other Centers, FRC was to prove itself a giant contributor to the advancement of aerospace technology.

During this period, FRC operated numerous experimental aircraft for the exploration of supersonic flight. The NACA, and later NASA, worked with civilian aerospace companies and the military services to develop and test new airfoils, fuselage shapes, and propulsion systems. They included rockets, jets, reaction control systems, delta and straight wings, variable-geometry airfoils, and semi-tailless configurations. Manned flight had just barely exceeded speeds of Mach 3 and altitudes in excess of 126,000 feet.

In June 1959, the rocket-powered North American Aviation X-15 arrived at Edwards AFB. Destined to become one of the most famous "X-Series" aircraft, it would further extend man's reach into the edge of space and the hypersonic speed regime. Eight out of the 12 pilots in the program qualified for astronaut wings in the X-15. They were not the first humans to leave the Earth's protective envelope of air. That honor went first to Soviet cosmonauts and American astronauts riding in ballistic capsules in early 1961. The X-15 pilots were the first to leave the atmosphere in a winged aircraft, controlled by a pilot, beginning with Maj. Robert White in July 1962. White had already made a name for himself as the first human to fly a winged aircraft four, five, and six times the speed of sound! Aerospace research had entered an exciting new realm and the NASA Flight Research Center was on the cutting edge. It was into this world that I arrived on 23 February 1963.

NASA photo E65-00644

The NASA Flight Research Center (later named in honor of Hugh L. Dryden) was built adjacent to Rogers Dry Lake at Edwards AFB, California. Small in size compared to other centers, it was nevertheless a giant contributor to the advancement of aerospace technology.

NASA Edwards Flight Operations

When I reported for duty at FRC, there were six pilots on staff. Neil Armstrong had recently transferred to the Astronaut Office in Houston, Texas, so I filled his slot. Fred Haise joined the team shortly after I arrived.

Joe Walker, the chief pilot, was well known throughout the flight test community because of his work in such research aircraft as the D-558-I, D-558-II, X-1E, X-3, X-4, X-5, and X-15. He also served as a research pilot on flight programs involving modified and unmodified production aircraft including the F-100, F-101, F-102, F-104, and B-47. He had received his early flight training and experience as an Army Air Corps pilot during World War II, flying many missions in the twin-tailed Lockheed P-38 Lightning. For his war service, Joe received the Distinguished Flying Cross and the Air Medal with seven Oak Leaf Clusters. A native of Pennsylvania, like myself, Joe joined the NACA in 1945 as a physicist at the Lewis Propulsion Research Laboratory in Cleveland, Ohio. He transferred to the High Speed Flight Research Station

(HSFRS) at Edwards in 1951 and had since been honored numerous times for his achievements.

Joe was highly respected in the pilot's office and, although he did not rule with an "iron hand," he was a strong leader. He expected every pilot on his team to do a professional job. He led by example and expected others to live up to his standard. If Joe ever had a problem with another pilot, he brought it up immediately and aired it out in order to expedite a solution. On occasion, he exhibited a quick temper, but it subsided quickly. He never held a grudge or brought up past problems. He always looked ahead and kept the team moving forward. I soon developed a great respect for Joe that only grew over the years.

John "Jack" McKay joined the NACA as a model builder at Langley in 1940. The son of a Navy officer, he became a Naval aviator himself during World War II. After the war, he earned a degree in aeronautical engineering and returned to the NACA. Initially joining the HSFRS staff as an aeronautical research engineer in February 1951, he transferred to Flight Operations in July 1952. While there, he specialized in high-speed flight research in the D-558-I, D-558-II, X-1B, X-1E, X-5, F-100, F-102, F-104, and F-107. He was assigned to the X-15 project in 1958. Jack had not participated in the number of "record setting" flights that Joe Walker had, but flew many productive research flights. It was the sort of "pick and shovel work" required for completing a set of data for an aircraft's flight envelope. The pilots who perform such tasks don't get nearly the credit they deserve.

Jack was also known for his ability to handle emergencies, and he had more than his share. While in the Navy, he survived crashing a Grumman F6F in shallow water due to a blunder by his flight leader. In 1955, Jack co-piloted a B-29 mothership for the X-1A when a liquid oxygen tank in the X-1A exploded. Jack helped jettison the stricken research airplane over a bombing range and assisted the pilot in making a safe emergency landing. The following year, Jack was in the D-558-II awaiting drop from the P2B-1S mothership when trouble struck. Having problems of his own, Jack had just called "No drop!" when the P2B suffered a runaway propeller. Recognizing the danger, the mothership crew jettisoned the D-558-II, leaving Jack to dump fuel and make an emergency landing on the lakebed. Moments after drop, the propeller broke apart, slicing through the fuselage of the mothership. It would have killed Jack if he had not been jettisoned. In November 1962, a crash landing of the X-15 left him with serious injuries. He recovered and returned to flying status, but he suffered discomfort for the rest of his life.

I found Jack to be a very colorful and likeable character. He and his twin brother Jim were two fine Scotsmen that I had been warned about while at Langley. Tommy Brady, my former Langley crew chief was a fellow Scotsman and friend of the McKays. He warned me never to try to stay even with the McKay brothers when it came to drinking. A few years later, I learned that Tommy was serious and accurate in his warning. Jim McKay was a research engineer at Langley while I was there. He helped secure Jack a position at the HSFRS and later transferred to Edwards himself. It was nice to know someone such as Jim in my new environment.

Stanley Butchart, yet another World War II naval aviator, had flown a number of combat missions in the Pacific. As pilot of a Grumman Avenger torpedo plane, Stan earned the Distinguished Flying Cross and a Presidential Unit Citation. While flying in the Naval Reserve after the war, he earned degrees in aeronautical and mechanical engineering. Following graduation in 1950, Stan worked as a design engineer for Boeing aircraft. He joined the NACA HSFRS as a research pilot in May 1951. Stan was involved in research involving stability and control, performance, and handling qualities of such aircraft as the D-558-I, D-558-II, X-4, X-5, F-86, F-100, and F-102. He also flew "heavy" transports including the KC-135, CV-880, and CV-990 for handling qualities and turbulence data. Stan, who worked on the B-47 Body Design Group at Boeing, flew the Station's JB-47A for several years. The aircraft was used for stability and control tests, aerodynamic noise investigations, approach and landing surveys, gust loads studies, handling qualities evaluation, and development of the X-15 High Range. He also flew B-29 and P2B-1S motherships for the rocket research aircraft and the Center's C-47 transport. When I arrived in March 1963, I was assigned to fly a modified JetStar aircraft with Stan. He and I began a friendship that has lasted several decades.

Milt Thompson served as a Naval aviator during World War II and later earned a degree in engineering. While flying in a Naval Reserve squadron in Seattle, Washington, he worked for Boeing as a structural test engineer and flight test engineer. Milt joined the NACA in 1956 as an engineer at what was then known as the High Speed Flight Station. In January 1958, he joined the pilot's office and flew such aircraft as the F-51, T-33, F-100, F-102, F-104, and F-105. In 1962, Milt was selected by the Air Force as the only civilian pilot scheduled to fly the X-20 Dyna-Soar space plane. He had served as a consultant to the program since 1959. The project, unfortunately, was canceled before the vehicle was even built. Subsequently, he joined the X-15 program. Milt also worked on the Paraglider Research Vehicle, known as Paresev. It looked kind of like a tricycle under a triangular "hang glider" parasail. Besides the X-15, Milt is best remembered for his work with the wingless lifting body research vehicles. These half-cone shaped craft were sometimes called "flying bath tubs." When I arrived, Milt had just started tow tests with a wooden lifting body called the M2-F1. He was an excellent pilot. I was very surprised, some years later, when Milt decided to hang up his pilot's hat and move back into engineering work. He decided to make the move while he was still fairly young rather than wait until he was too old to fly and perhaps too old to start a new career path.

Bill Dana received a Bachelor of Science degree from the U.S. Military Academy in 1952 and served for four years as a U.S. Air Force pilot. After earning a Master of Science degree in aeronautical engineering, he interviewed for a job with the NACA in September 1958. As it turned out, he started his job as a stability and control engineer at Edwards on the day that the NACA became NASA. On 9 September 1959, Bill transferred to Flight Operations as a research pilot. He was involved in high-speed jet research when I arrived at Edwards, and he was destined to join the X-15 program in a couple of years. Bill also made numerous flights in the lifting body vehicles and many other aircraft.

Bruce Peterson worked as an aircraft assembler at Douglas Aircraft Company from 1950 to 1953. He served as a Marine Corps pilot for several years and earned a degree in aeronautical engineering. He also remained active with the Marine Corps Air Reserve, flying A-4 Skyhawks. Bruce joined NASA in August 1960 as an engineer at the Flight Research Center. He joined Flight Operations in March 1962 and was initially assigned to the Paresev program. Bruce later flew various lifting body vehicles. He was injured in a crash in 1967, while landing the M2-F2 lifting body. Bruce was hospitalized for some time and lost an eye due to a secondary infection while in the hospital. He continued to fly and later transferred into engineering and flight safety and continued a very distinguished career.

Fred Haise became a Naval Aviation Cadet in 1952. He completed Navy flight training in 1954 and served as a Marine Corps fighter pilot until September 1956. While serving in the Oklahoma Air National Guard from 1957 to 1959, Fred earned a degree in aeronautical engineering from the University of Oklahoma. In September 1959, he went to work as a research pilot at the NASA Lewis Research Center near Cleveland, Ohio. Although he remained on staff there until March 1963, he was recalled to service as a fighter pilot with the U.S. Air Force 164th Tactical Fighter Squadron at Mansfield, Ohio from 1961 to 1962. He then transferred to FRC shortly after I did.

These were the pilots I worked with in the NASA FRC Flight Operations office. I was proud to be a part of this talented team and contribute to its mission.

Moving West

The logistics of moving my family to Edwards was not a small one. First, we had to sell our home in Virginia. I did not want to leave town with our house in the hands of a realtor because it would have been much more difficult to conduct the sale from California. More importantly, we needed to get as much of our equity as possible out of the Virginia property in order to facilitate the purchase of a new home in the Antelope Valley. We decided to set the price about 20 percent below market value to expedite the sale. That worked. We were left a little shorter on equity than we desired, but at least we did not have the problem of selling from 3,000 miles away.

Audrey and I also decided to swing by Pittsburgh to visit our parents on the way west. Audrey planned to remain in Pittsburgh while I drove to California

High Desert Flight Research

to locate a place for the family, and then she and the three children would fly out to join me. This seemed like an excellent plan until I arrived at FRC. There, I discovered that the pilot's office had attempted to contact me to tell me to go directly to Marietta, Georgia, for checkout in the Lockheed JetStar aircraft. I had departed Langley before they could contact me and they had to wait for me to show up to deliver the news.

I barely had time to locate a rental home, pick Audrey and the kids up at the airport, get the furniture moved in and catch my breath. In almost the next breath, I had to tell Audrey that about four days after her arrival, I was leaving for Marietta, Georgia, for a few weeks of training. This was a little difficult for her to take because one of my main arguments for the transfer to Edwards was that I wouldn't have to travel as much as I had at Langley. During my final year at Langley, I spent one-third of the time away from home on support trips. Fortunately, this was not the norm at FRC. It just amounted to bad timing in this case. I moved my wife and three children, Sandra, Karl, and David into a rental home with a garage stacked with unpacked household goods. I then bid them adieu and went to Marietta.

I was assigned to the JetStar training school with Stan Butchart. The training took place at Lockheed's Marietta facility from 13 March through 29 March. The JetStar was not that big, but it was a multi-engine aircraft with all the systems and problems of the larger aircraft. Since the JetStar required two pilots, I was a natural to join Stan for this project.

Stan and I got along very well from the very start. We had both gone through the Naval Air Training Command, although I had trailed Stan by about ten years. There seemed to be a little extra communication between us and I suspect that a lot of that was due to our similar backgrounds. Naval training developed a philosophy toward flying that stays with the trainee the rest of his life. I'm sure that is true with the Air Force training, too. It is such an intense experience that it would seem impossible not to retain attitudes and philosophies gathered there.

The time passed quickly for me because it was a worthwhile school and the JetStar was great to fly. It was my first experience in training at a school designed for commercial operators who flew the aircraft in an airline environment. I also had to memorize numerous emergency procedurs.

When I returned to Edwards and the office routine, Audrey was glad to have me back and get some support on the home front. She had done well with the kids in a strange town and strange neighborhood with (as she noted) some strange people. The strange people that she referred to were some of our neighbors that walked outside during a rainy day and seemed to greatly enjoy the downpour. Being from Virginia where she had too much rain, Audrey thought this very unusual. After we had been in the desert a few years, we too, walked out to enjoy the rain when it came to the desert.

Smell of Kerosene

NASA photo E63-10201

Stan Butchart and Don Mallick delivered the Lockheed JetStar to NASA Flight Research Center in May 1963. The aircraft was modified to serve as a General Purpose Airborne Simulator (GPAS).

My first year at Edwards was exciting. I started flying the Aero Commander and the C-47 in February. These aircraft were used to transport personnel between FRC and various locations in the western United States. I often flew these aircraft to Los Angeles, Burbank, and Ames Research Center. Occasionally, I flew these planes to Beatty, Ely, and Tonopah, Nevada, to support the High Range tracking facilities for the X-15. During the course of the year, I accumulated over 200 hours flying time in 12 different types of aircraft. It looked as if I had made the right decision regarding the transfer. Some of the various aircraft types included the JetStar (also called a C-140 or L-1329), C-47 Skytrain (we called it the "Gooney Bird"), T-33 Shooting Star, T-37, F5D-1 Skylancer, A-5 Vigilante, F-100 Super Sabre, M2-F1 lifting body, Aero Commander, two Schweizer sailplanes, and the double-sonic F-104. The Skylancer was unusual, as only a few prototypes were built. It was designed as a follow-on to the Navy F4D Skyray production fighter. As completely powerless aircraft, the sailplanes were an entirely new experience for me. My experience in the M2-F1 was not only powerless, but wingless as well!

Stan and I went back to Marietta on 22 April. We brought the JetStar back with us to FRC on 8 May. About six months after NASA accepted it from Lockheed, technicians began to modify the aircraft as an airborne simulator that could duplicate the flight characteristics of almost any other aircraft. A number of smaller aircraft (such as the JF-100C and NT-33A) had their flight control systems modified for similar work, but finding adequate space to put all of the computer equipment, servos, and associated equipment required for the variable-stability aircraft was a problem. The JetStar had ample space for the equipment and excellent flight performance characteristics.

We collected baseline data on the flying qualities of the aircraft and then sent it to Cornell Aeronautical Laboratory in Buffalo, New York. At the time, Cornell was the most experienced organization with regard to variable stability aircraft. It took more than a year to complete the modifications.

The modified JetStar was called the General Purpose Airborne Simulator (GPAS). It was a unique research tool that served us well for many years. As technology evolved, additional modes and improvements were added. The system was referred to as a Model Following System and was several steps beyond other computer-based simulators then in use. A specific aircraft's equations of motion could be programmed into the JetStar's system, which would then respond to control inputs just as the subject aircraft would. This could be accomplished even before the subject aircraft was built by using the equations of motion developed from ground-based simulators, wind-tunnel data, and mathematical predictions.

We found that the GPAS results were reliable and repeatable. When a set of equations was programmed and verified, the response and performance of the system were identical to simulations that had been completed during previous tests. Older simulations with "rate feedback" logic often changed from day to day, even with the same inputs or settings. They required a more involved verification checkout prior to being used for gathering data. Additionally, rate feedback systems did not have as wide a capability for simulation as the Model Following System. I consider the GPAS one of the most successful airborne simulators of all time. It served as a research workhorse at FRC for over two decades. Besides its GPAS function, the aircraft also provided data on laminar flow, propfan characteristics, ride qualities, and the Microwave Scanning Beam Landing System for the Space Shuttle.

Variable Stability F-100 Super Sabre

While the JetStar was undergoing modification, Joe Walker assigned me to the JF-100C as prime research pilot. The aircraft was equipped with a modified flight control system for variable-stability research. Bill Dana had been prime pilot in the JF-100C, but his assignment to the X-15 program dominated his workload. My experience flying the JF-100C at Langley made me a natural to pick up the program. In the JF-100C, the pilot used a side-stick

NASA photo EC62-00144

This JF-100C was modified for variable stability and equipped with a side-stick controller. It was flown to evaluate minimum levels of controllability.

controller to select the variable stability system once the aircraft was airborne at a safe altitude. Thus, in the event of any uncommanded control input or other problem with the experimental system, the pilot had sufficient altitude to recover or eject.

I normally climbed to a specific altitude and airspeed before engaging the system. Once engaged, I flew the aircraft through a series of maneuvers to evaluate the system. It required an intense effort to complete them smoothly and efficiently. I climbed to high altitude for a margin of safety. This provided the opportunity to evaluate extremely poor handling characteristics to the point of loss of control while still having enough altitude left to recover. Once the aircraft's control characteristics became marginal or divergent, the pilot disconnected the system and the aircraft reverted to the basic F-100 controls. The JF-100C provided an airborne simulation to evaluate and plot envelopes of controllability. This helped aircraft designers find the minimum level of controllability that a pilot could stand and still fly the aircraft safely and complete his mission. Minimum aircraft stability, control power and damping levels could be determined from the data.

The JF-100C came to Edwards from Ames Research Center where it was converted for three-axis variable stability simulation in October 1959. Bill Dana and several other pilots had used the JF-100C for variable stability research between March 1961 and May 1963. Toward the end, however, the aircraft had been frequently used for X-15 support missions. I made my initial check flight in the JF-100C on 16 May 1963 and was soon engaged in an intensive evaluation program. The first time I activated the variable stability system at 35,000 feet, I encountered a very interesting phenomenon. As I performed evaluation maneuvers with the side stick, I noted a slow reduction in control power (response to pilot inputs). This continued until there was no control response at all from the test system. Control power then began to return, but the pitch and roll response had reversed. Normally, when a pilot wants a "nose-down" response, he pushes forward on the stick. When he wants "nose-up" he pulls back on the stick. With this glitch in the variable stability system, everything was "ass backward" to what the pilot was accustomed.

After assuring myself that I had command of the aircraft, despite the reversed controls, I tried to fly the evaluation task maneuvers. I was amazed that, even with a reversed control system, I could still complete my tasks with some degree of precision. I determined that once the aircraft excursions became very large or rapid (which increased the flying task greatly) there was a natural tendency to return to my normal response habits. In this case, that would drive the aircraft out of control or would disconnect the test system, reverting to the basic F-100 controls. When I returned to base, I wrote up a "squawk" report on the system. A maintenance inspection revealed that the test electronics that were supposedly installed in a pressurized container were in fact not adequately pressurized. At high altitude, the container leaked and the lower air pressure resulted in a breakdown of the electronics, and thus the reversed controls.

The JF-100C also had a problem involving a severe nosewheel shimmy during landing. I encountered this on my very first flight, and it surprised me because I had never had this problem with our F-100 at Langley. When I approached Bill Dana about the problem, he said, "Yeah, its been doing that for some time." "Don't sweat it," Bill added, "you Langley pilots are too touchy and sensitive." It was a discouraging response. I later found out that the maintenance crew had been trying to solve the problem. After each flight, they checked the nosegear steering, dampers, and whatever else they could, to try and find the answer, but it seemed to elude them.

About three flights later, I experienced the most aggravated nosewheel shimmy that I had ever encountered. I thought the front of the aircraft was going to come off during the landing rollout. When I reached the parking area and shut the engine down, I told the crew chief that I was not going to fly the airplane again until some definite solution was found. This was sort of a hard line to take with the people who strive to make the aircraft as safe as possible for each flight, but I was ready for a change. My verbal and written comments that the aircraft was not flyable were reinforced by the devastation that the crew found when they opened the nose section of the aircraft. It was a complete shambles. Tape recorders had broken from their racks and other instrumentation was wrecked. It was now obvious to the maintenance crew that the aircraft shimmied too much to fly.

The F-100 was down for an extended time to repair the instrumentation and, hopefully, the nosewheel problem. The system was again thoroughly checked out. In addition, almost as an afterthought, the main landing gear was inspected thoroughly. Lo and behold, it was discovered that the main gear trunnions (the bearing pivot points where the main gear is attached to the airframe) were extremely worn and loose. After many hours of maintenance, they were replaced. On the checkout flight, the landing rollout was smooth and comfortable. It surprised me that it had taken two years to fix the problem.

F-104 Starfighter

During the remainder of 1963, I settled into my new environment and began to expand my responsibilities. Like the earlier stages of my flying career, each step was a new challenge, more advanced than the last. This was good because it prevented me from becoming stagnant in one area. It made life interesting and I never grew bored with my work. In fact, I am reluctant to call it work at all. Along with the variable stability F-100 and the JetStar, I flew chase and support flights for other test programs. This allowed me to become familiar with each program at the Center.

On 24 May, a sunny Friday, Bruce Peterson checked me out in the F-104B Starfighter. The F-104 was the workhorse of the Center's support aircraft. We had two of them when I arrived. Another had been returned to the Air Force in 1961 and one was lost when Milt Thompson had to eject due to a flap malfunction in 1962. At FRC, the F-104 served as a safety chase for high-speed programs and provided range support for the X-15. It was also used for pilot proficiency flights.

Many people may not be aware that actual research flights are not as frequent as pilots might like. There are hours and hours of ground preparation required for testing special systems and instrumentation. If a pilot's test aircraft is "down" (not flyable) for two to three months, he needs to fly other aircraft in order to maintain his skills. The F-104 was that aircraft and I enjoyed flying it whenever I got the chance. Shaped like a dagger, the fuselage seemed as if it had been wrapped tightly around the engine. Aerodynamic surfaces consisted of stubby, razor-sharp wings and a T-tail. I had to adhere strictly to the operating limitations, but the Starfighter had very nice handling qualities. The performance was excellent too. It had a top speed of Mach 2. The F-104 also provided the opportunity to experience a wide variety of drag conditions from very low (with gear and flaps up) to very high (with gear and flaps down). This allowed the F-104 to serve as high-speed or low-speed chase. The ability

NASA photo E66-15127

NASA pilot Bruce Peterson checked Mallick out in the Lockheed F-104B Starfighter. The airplane served as a research platform and was also used for mission support and proficiency flights.

to make low lift-over-drag (L/D) approaches made the F-104 the perfect aircraft to chase the X-15 or lifting bodies during their steep power-off approaches and landings.

Later in the year, the Center received three single-seat F-104N aircraft from Lockheed. They were purpose-built for NASA and served as support aircraft for many years. Lockheed built them to the same specifications as the F-104G, but without any armament since they were intended for civilian use.

Probing the Edge of Space

My first exposure to rocket-powered aircraft was during one of Joe Walker's flights in the X-15. At the time, he and Jack McKay were the NASA pilots on the project. Major Robert White was the Air Force project pilot and he ultimately became the first human to exceed Mach 4, Mach 5, and Mach 6. Joint programs such as the X-15 included personnel from several different agencies. The Navy also provided some funds and support for the X-15 and their pilot was Lieutenant Commander Forrest "Pete" Petersen. NASA was assigned "operational control" of the program and the vehicles were flown from NASA

Smell of Kerosene

facilities and data gathered by NASA technicians. The Air Force provided some chase aircraft and pilots, radar tracking and logistic support, and the launch aircraft and crews.

This was a new situation for me. At Langley, our operation primarily involved NASA personnel and equipment, although we operated from the Air Force facilities at Langley AFB. There were times at Langley when we invited other test pilots from the services or industry to fly an aircraft or participate in a program, but these were not joint programs like those at FRC. I was very impressed with the smoothness of joint operations at Edwards. Seldom were there any conflicts, and the USAF personnel assigned to the projects were always top-notch people who seemed to like their assignments. I had worked there six months before I realized one of the X-15 engineers, Johnny Armstrong, belonged to the USAF and not NASA. Johnny was so involved, and fit right in with the NASA system, that I didn't realize he was from "the other side of the house" at Edwards. That was just the kind of teamwork we had.

As the Center's major program at this time, the X-15 required a great deal of logistics and support. The program personnel showed their dedication by putting in overtime whenever necessary. There were a number of occasions when, during the late afternoon on a day prior to a scheduled morning X-15 flight, a hydraulic sample from the vehicle's control system appeared questionable. I would then stay after normal working hours and fly the chief X-15 engineer, Ron Waite, and a sample of the fluid to the North American Aviation laboratory in Los Angeles for a precision analysis. Sometimes we got back late at night with the sample approved and then got up early the next morning to launch the flight. Such efforts required extra work, but it was never onerous. We accepted it as part of the operation. Job satisfaction was our chief reward whenever "going the extra mile" helped launch a successful flight. The X-15 program was complex and challenging. Each participant, no matter how small his or her role, experienced immense pride of accomplishment. We were truly a team. Such cohesiveness and pride were hard to duplicate in later programs.

Early in 1963, before I started flying weather or chase flights for the X-15, operations engineer Bill Albrecht asked me if I would like to join him on a trip to the lakebed. We were to place smoke markers along the X-15 runway to provide a visual reference for the pilot and then remain out there to observe the landing. Normally, the X-15 landed on Runway 18 near the north end of Rogers Dry Lake. The lakebed stretched nearly 11 miles from north to south and almost 5 miles east and west. Air Force personnel had used thick, dark oil to mark a number of runways on its surface. The runways were about 300 feet wide and between two and seven miles in length. Runway 18 was nearly four miles long.

For this flight, the X-15 launched from beneath the wing of a B-52 mothership about 50 miles north of Las Vegas, Nevada. An Air Force asset,

NASA photo EC88-00180

An X-15 rocket plane (right) glides to touchdown following a research flight accompanied by an F-104 chase plane.

the B-52 was commanded this day by Maj. Fitz Fulton, a pilot that I would come to know better in a few years. After launch, Joe Walker piloted the X-15 to an altitude above 300,000 feet and achieved a speed over Mach 4 while completing a number of precision research test points. Now, with its rocket fuel depleted and in a decelerating glide back to Edwards, the pilot was about to bring it home. The X-15 normally landed to the south after passing over the community of North Edwards and Highway 58. About 30 minutes prior to the X-15 landing, a radio vehicle was dispatched to the north end of the runway. As the X-15 approached Edwards, several smoke bombs were ignited along the side the runway to give the X-15 pilot information on wind conditions along with an aim point for his final glide to touchdown.

I watched as a little black dot passed overhead, turned nearly in a circle, and dropped toward the lakebed. As the X-15 began its three-mile final approach, it displayed its upper surface during a steep descent. As it leveled off toward us, it again became a small black dot that slowly grew in size as it silently approached the end of the runway. It appeared to be aimed directly at us even though we were offset a hundred feet from the runway. The rocket plane rushed toward us, seeming to jump from a distant, silent black spot to full-size in an instant. Suddenly, it sounded like a roaring freight train with the rush of air flowing over the sharp edges of its wings and tail surfaces. As it passed abeam our position, the nose wheel gear and landing skids extended and the aircraft grew small again as it touched down on the lakebed about a mile south of our position. A telltale trail of dust arose as it slid to a stop.

I was so taken aback by the experience that I just looked over at Bill with astonishment on my face. "Really something, isn't it?" he said with a grin. Bill had spent a great deal of time putting out smoke markers for landings, and he took great pleasure in seeing the awestruck expressions of visitors such as myself. Yes, it was really something. Recovery convoy vehicles sped toward the X-15, now stationary on the hard clay. As I watched the distant black speck through the settling dust, I wondered if I would ever have the opportunity to fly that machine. Unfortunately, I never did. It was a matter of being in the right place at the right time. I was in the right place, but my timing was off. I felt a personal loss, not being able to fly it, but I had many other opportunities ahead of me.

I had great respect for Joe Walker and observing him in action always enhanced that feeling. On another X-15 flight operation, I flew an F-104 uprange to the launch location north of Las Vegas for a weather check. The X-15 was always launched in an area that had a dry lakebed nearby. In the event that a post-launch problem prevented the X-15 from making the flight back to Edwards, the pilot could use a contingency lakebed for an emergency landing. Therefore, we needed acceptable weather at the launch lake and intermediate lakes as well as at Edwards. In this case, "acceptable" meant that the X-15 pilot could see his emergency landing lake from under the wing of the B-52 mother ship and have adequate visibility to make a visual approach to that lakebed if the need arose.

On this particular morning, I flew the two-seat F-104B. As this was my first weather flight, Joe Walker rode in the rear seat to coach me on interpreting the conditions we encountered. As an experienced X-15 pilot, Joe was the best instructor that a fellow could want. The weather observers call was very important in that it could hold or scrub (cancel) the mission. Preparations for an X-15 flight were extensive, and it was no small thing to cancel a launch for any reason. In fact, every group involved strove to keep its part, system, or operation in an "A-OK" status so as not to be the reason for canceling a flight. It was a matter of pride and dedication.

As Joe and I flew into the Las Vegas area, we found the weather terrible. To me this was an easy call. I wasn't too happy about flying the F-104 in this weather, let alone call for the launch of an X-15. As my instructor, Joe asked me my opinion of the weather. I was surprised that he even bothered to ask. "Hell, Joe," I said, "I wouldn't launch my worst enemy in this crap." Joe just laughed. "Right," he said, "but don't tell them on the radio yet." I was really confused when Joe called NASA Flight Operations and told them that we were still looking at the weather and would let them know our call in about 10 minutes. About five minutes later, we received a call from Operations saying that the mission had been scrubbed due to a maintenance problem. Joe laughed again and said, "I knew those guys had problems with the inertial navigation system this morning and I wasn't going to let them off the hook with a weather cancellation. I wanted them to fix it or own up to it." That was Joe Walker. He knew the system completely and he made it perform to the maximum.

Besides flying weather check missions and making launch/no-launch calls, I also flew chase for X-15 flights. Some of the chase assignments were of a stand-by nature where I flew the F-104 in an area where I could assist the X-15 pilot in an emergency or deviation from the normal flight plan. Others were launch chase missions where I flew beside the mothership B-52 up to the point where the X-15 fired its engine and climbed away. During launch chase I observed control motions and auxiliary power unit (APU) performance as the X-15 pilot prepared for launch.

I will never forget my first launch chase. The X-15, operating in igniter idle (the pre-start condition of the rocket motor), dropped off the hooks from the B-52. A few seconds later, the pilot fired the powerful XLR99 rocket engine, capable of developing over 57,000 pounds of thrust. As the X-15 dropped a few hundred feet below the launch plane, a giant white vapor plume streamed from its tail. With a tremendous burst of acceleration, it surged forward and climbed away from the B-52 and the chase jets as if we were standing still. The sensation made me glance down to check my own airspeed to make sure my aircraft had not just stopped in flight. Actually, I was still flying at over 500 miles per hour!

Not surprisingly, with an aircraft as sophisticated and complex as the X-15, there were frustrations and cancellations. Some of the most bitter disappointments came when a flight canceled just seconds before launch because some pressure or voltage was too high or low, or a piece of equipment malfunctioned. Such cancellations, or "scrubs," took the wind out of our sails. Eventually, however, technicians fixed the problem, managers rescheduled the flight, and we started over again. Conversely, once a flight had been successfully completed, there was a great feeling of accomplishment and relief. Then we threw a post-flight party. This gave the team members a chance to celebrate and let off a little steam.

I recall one aborted flight when Joe Walker was the pilot and the flight controllers called for abort just seconds before launch. Apparently, engineers in the control room had been watching and trying to fix a problem as we approached the launch point. They worked on it without success until they ran out of time. I was flying one of several chase aircraft. There were three chase planes in the vicinity at the time: a launch chase, a photo chase (me), and another to cover a launch lake landing if required.

Walker was a true gentleman and if he had any problem at all, it was that sometimes his anger would flare up. A late scrub was a difficult situation for all involved. Immediately after the scrub was called, communications stopped. There was a deadly silence as the ground controller stopped reading the launch checklist. I think we were all waiting for Joe to let off a little steam, but he didn't. He just spoke calmly in his western Pennsylvania drawl: "You don't all have to clam up on me just because we aborted." Again, there was a long pause indicating that personnel on the communications circuit were still reluctant to comment. I finally keyed my mike, without identifying myself, and said, "Joe, if you think you have problems, you should hear about my brother in Pittsburgh." Again, there was a very pregnant pause on the airways and Joe finally broke down and said in a voice so low that you could barely hear it, "What was that?" I think he knew when he responded that he had opened the door, but he couldn't hold it back.

"Well," I answered again without any identification, "my brother has a wife, a girlfriend, and a note at the bank, and all three of them are overdue." Again, the silence, but I learned later that it went over well in the control room and other support stations. It was just a way to relieve tension. Apparently, nobody recognized my voice. When I returned to the pilot's office after landing, my secretary asked me if I had heard that joke and wondered who might have said it. I acknowledged that I had heard it, and said it sounded like Bill Dana to me. It wasn't unusual to point the finger at some innocent soul in a case like this, all in good fun, of course.

The X-15 project was one of the most successful high-performance flight test programs ever conducted. The first flight occurred in June 1959 and the last flight was on October 24, 1968. North American Aviation built three X-15 aircraft, one of which later received modifications for increased performance. Considering that these were rocket-powered flights into an unexplored environment and performance envelope, it is amazing that only one aircraft and one pilot were lost during a program that spanned nearly 10 years and 199 flights. The X-15 carried pilots to the edge of space at fantastic speeds, while serving as a flying laboratory. It was truly a remarkable program.

Lifting Body, Paresev, and Test Pilot School

Flying and working at Edwards was everything that I had hoped for, and my prospects for the future seemed unlimited. I continued building flight time and experience in the F-104 and the more I flew the aircraft, the more I liked it. My prior C-47 flying experience from NASA Langley allowed me to participate in a new area of research, the Lifting Body Program. The first lifting body research vehicles resembled flying bathtubs. They had no wings, but generated lift from their shape, which was a blunted half cone.

The first lifting body research vehicle was constructed mostly of plywood, with ribs and bulkheads like a boat. William G. "Gus" Briegleb of Sailplane

Corporation of America signed a contract with NASA to build the plywood shell for around $10,000. The "lightweight" lifting body aircraft, designated M2-F1, weighed less than a thousand pounds and was unpowered. During initial flight tests, a Pontiac Catalina convertible automobile that had been reconfigured as a ground-tow vehicle pulled the M2-F1 across the lakebed until it reached takeoff speed. The Pontiac was capable of about 130 miles per hour by itself and could tow the lightweight lifting body up to about 100 miles per hour. At about 85 to 90 mph, the M2-F1 lifted off the lakebed and flew under tow behind the Pontiac. With the expansive dry lakebed available, the pilot had several minutes for test and evaluation behind the Pontiac before he released the towline and returned to earth. Once these tests proved that the M2-F1 developed enough lift to fly and was controllable, the next step was to tow the M2-F1 to high altitude behind an airplane. The C-47 seemed to be the best aircraft for the job and I was selected as one of the tow pilots.

The M2-F1 was attached to the C-47 by a releasable towline. The C-47 then accelerated to its normal takeoff speed, but held it on the dry lakebed with forward elevator until the M2-F1 pilot had lifted off and climbed to a position above the C-47 in order to clear its wake. At this time, the tow pilot released forward elevator and allowed the C-47 to become airborne. The joined aircraft then began a climb to around 13,000 feet on a pre-determined track toward the launch point. After release, the pilot flew a series of test points and evaluations, culminating in a deadstick landing on the lakebed.

Radio communications between the C-47 pilot and the M2-F1 pilot left something to be desired. Engine noise and other factors made it difficult to hear each other. The C-47 had an observer that watched the M2-F1 at all times through the bubble sextant dome on top of the fuselage, and he kept the tow pilot informed about the condition of the lifting body.

One of the most undesirable flight characteristics of the M2-F1, and other lifting body aircraft that followed, was the vehicle's sensitivity to sideslip. A small amount of rudder input resulted in a severe roll reaction. This was further complicated by the fact that the vehicle's roll control system was not powerful enough to counter the roll, and the pilot could very easily encounter a tremendous "Dutch Roll" instability. On two different occasions, Air Force test pilot Jerry Gentry encountered this instability shortly after the C-47 had taken off. He actually completed a slow roll on the towline behind the C-47 before he could pull the release. "I'm the first one to roll this toad," Jerry said after making a safe landing on the lakebed. Jerry went on to fly the heavyweight rocket-powered lifting body test vehicles with no problems.

During this time, FRC was also involved with another unique research program involving the Paraglider Research Vehicle (Paresev). Kite-parachute studies by NACA Langley engineer Francis M. Rogallo spawned a triangular wing formed from several long spars with fabric in between. Forward velocity allowed air to fill out the shape of the wing as it pushed upward on the flexible cloth. This wing developed reasonable lift and could be controlled by shifting the location of the load or vehicle attached below, and also applying inputs to the wing itself via control lines.

NASA engineers evaluated the "Rogallo Wing" as a possible recovery system for spacecraft returning to Earth after reentry from orbit. This system was viewed as an alternative to parachuting returning spacecraft into the ocean. It offered greater crossrange, controllability, and the option of landing on a runway or lakebed. Two cloth parawings of various sizes, and one inflatable version, were alternately mounted on a framework above a simple aluminum tube tricycle with a pilot's seat and controls. The Paresev was built "in-house" at FRC for a mere $4,280 in construction and materials. Several pilots from FRC, Langley, and the Manned Spacecraft Center in Houston, Texas made flights in the Paresev. The vehicle was towed behind a utility vehicle and various light aircraft, then released in a similar manner to the M2-F1. The Paresev proved challenging to the pilot and difficult to handle.

North American Aviation built a larger paraglider with a mock-up Gemini-type capsule and a Rogallo wing, but their project pilot sustained serious injuries during a landing mishap. Questions arose regarding deployment of such a device from a spacecraft. Plans to apply the parawing recovery system to the Gemini spacecraft were dropped. Program delays, poor flight test results, and cost factors all contributed to cancellation of the project. This did not mean that Paresev was a failure. Learning what doesn't work is as important as learning what does work. Research moved on to other, more fruitful, areas. That meant the lifting bodies, which represent an evolutionary step toward the Space Shuttle.

In November 1963, Joe Walker approached me and asked, "Don, would you like to attend the USAF Test Pilot School?" There was no hesitation on my part. I gave Joe a solid "affirmative." The military test pilot schools were well known and highly respected by all in the flight test community. When I had been a senior in aeronautical engineering at the University of Florida, the Navy invited me to come back on active duty with a regular commission. I said that I would, only if they would also give me orders to the Navy Test Pilot School at Patuxent River, Maryland. The Navy couldn't do it because they were short of fleet pilots, not test pilots.

As a NASA pilot, the biggest obstacle for getting into these schools was money. Test Pilot School training is expensive due to the flight hours required as part of the training and the expense of operating the aircraft. Joe got NASA to approve the funding, so Fred Haise and I entered Class 64-A at the Air Force Test Pilot School (TPS) at Edwards.

I completed the test pilot portion of the curriculum in six months. This school was one of the best training experiences that I have had. Both academic and flying courses were concentrated and accelerated. I could count on working until midnight almost every evening during the academic portion of the program. The flight portion was highly regimented. Not only did I need to fly precisely, but also expeditiously in order to complete as many test points as possible per flight. The flight test cards always had an additional 20-percent of planned points beyond what normally could be accomplished. In addition, an alternate test card was always available in the event some problem interrupted the original test plan. The primary aircraft used by the school included the T-33, T-38, and B-57. In addition, students had the opportunity to fly the Cornell variable-stability NB-26H and NF-106B.

It was a fine school, and I found the training exceptional. I had flown with the NACA and NASA for six years before getting the opportunity to attend TPS, so I could evaluate the impact of test pilot school experience on my work at FRC. In my opinion, it would take a military pilot three years to accrue the same level of test experience provided by the TPS curriculum. In addition, the school provided a broad base of knowledge for the test pilot, compared to some organizations that concentrated in specific areas of test and evaluation.

The TPS commandant at this time was Col. Charles E. "Chuck" Yeager. I had met Chuck at NASA sometime earlier, and I liked the West Virginia hillbilly from the start. Chuck had grown up in the hills not far from where Joe Walker and I had lived in southwestern Pennsylvania. Chuck had a tremendous background in flying operational and test flights. I respected Chuck for his accomplishments, and it was obvious that students in the school seemed a little awed in his presence. We did not see too much of Chuck around the school because he was usually at the Pentagon taking care of the politics of the job. I'm not sure if Chuck ever flew with any of the students. At any rate, the school was staffed with outstanding instructor pilots.

I only had a few contacts with Chuck in my six months at the school. On one Thursday afternoon, normally set aside for athletics and physical activity, Fred Haise and I were reading photo panel film in the school's lab. Most of the military students in the class lived on base and could work evenings at the school, but Fred and I were both living in Lancaster and we took every opportunity to clean up the data reduction during the day. Chuck ran us out saying, "I don't care if you go to the bar and drink, but don't be in here on Thursday afternoons." So, Fred and I said, "To Hell with the data reduction." We took Chuck at his word and headed into Lancaster and stopped at our favorite watering hole for a beer or two.

Another occasion was when Chuck came into our classroom one morning in late January 1964 and said, "Come on, Mallick, we are going over to NASA." I followed him out and, as we drove over to FRC, I discovered that Chuck had arranged to get an air tow flight in the M2-F1 lifting body. He had also arranged for me to get a flight in it, and I'm sure that he coordinated this all with his old buddy, Paul Bikle, our director at NASA. Paul had been an engineer with the Air Force years earlier and had worked a lot with Chuck. Bikle had a great respect for Chuck's flight test ability.

Smell of Kerosene

NASA photo E63-10628

Milt Thompson (left) checks out Chuck Yeager (cockpit) in the M2-F1 lifting body as Bruce Peterson (secong from right) and Don Mallick (far right) look on.

Chuck and I had flown a number of M2-Fl ground tows behind the Pontiac before I started the Test Pilot School, so I wasn't surprised that we were now going to get some air tow experience. Chuck flew the lifting body first and then I flew it. It was quite an experience. Milt Thompson checked us out and he reminded me of the unusual sideslip characteristics. "Just keep your feet on the floor," Milt advised, "and off the rudders, and you won't get in trouble." I followed his advice, and it worked out fine. The takeoff was similar to the Pontiac tows in that there was a lot of dust kicked up off the lakebed. It was nice to reach rotation speed and climb out of the dust to a position slightly above the C-47 tow plane. It was a fun flight, in some ways similar to piloting a sailplane. I worked hard, keeping everything smooth and straight on the towline. When I released the towline, it was like breaking away from mother. I was on my own. I used just the aileron and elevators for control and stayed clear of the rudders. As we say in flight test, it was an uneventful flight. That meant there were no mishaps. The flare and landing, the most critical portion of the flight, went smoothly. It only took a few minutes to descend from a 13,000-foot launch altitude to touchdown on the lakebed. Without wings, lifting bodies come down rather rapidly, but they do fly and are controllable.

Chuck waited for my flight to end and we went back to the school together. That was the most contact I had with him during my time at the school, and it was probably more than any other student in my class had. I didn't realize it at the time, but I found out later that this day's events and my performance at TPS significantly influenced my future at NASA.

When I went back to class later in the morning, my fellow students asked where I had been. I don't know what they suspected, but when I told them what had happened they asked me, "Mallick, what the hell are you doing in this school?" "The same as you," I answered, "working my ass off and learning!" There were a lot of advantages to attending the Test Pilot School at Edwards as compared to the Naval Test Pilot School at Patuxent River, Maryland. For one thing, I could maintain my flight currency in the F-104 by getting over to NASA occasionally for a flight. I was also able to spend some time with my family, and I was able to participate in special the M2-F1 program. It was an incredible experience and I felt very fortunate.

I finished the USAF Test Pilot School in June 1964 and returned to FRC. Joe Walker called me into his office and began discussing projects that might be available for me. The ongoing X-15 program was a possibility that I was anxious to explore. The upcoming XB-70 program and the Lunar Landing Research Vehicle (LLRV) were also programs of great interest to me. I indicated that my preference would be the X-15 or the XB-70, followed by the LLRV. As it turned out, Joe started me in reverse order, and I'm sure the primary reason was that my prior background in VSTOL flying at Langley made me a prime candidate for the LLRV. Walker had recently retired from X-15 operations after flying it for a number of years and accomplishing a great deal. It was time to allow some of the other NASA test pilots to have a turn. I was destined for the LLRV and it was not my fate to fly the X-15. Joe had assigned himself to the LLRV with me, and it was an opportunity to work closely with him and get to know him better.

General Aviation Research

In NASA, it was not unusual to fly several test programs at the same time. This always made the work more interesting and avoided the occasional long wait between flights that sometimes occurred when I was flying just one test program. There were times, of course, when scheduling conflicts occurred. Such problems usually sorted themselves out as the highest priority program received the first attention.

Even as I began work with the LLRV, I was assigned to the General Aviation Test Program that FRC established to gather flight data for the manufacturers of civilian light aircraft. This data was also of interest to the FAA and other agencies. The unique thing about this program was that I was the program's chief pilot, chief engineer, chief of safety, and general all around honcho. The flight test instrumentation was self-contained and convenient to install in place

Smell of Kerosene

NASA photo EC70-02584

Mallick flew a variety of general aviation airplanes to gather information for manufacturers, the Federal Aviation Administration, and other agencies. The aircraft, including this Piper PA-23 Apache, were based at William J. Fox Field in Lancaster, California.

of a passenger seat. Naturally, some time was required to install the sensors on the aircraft. Once this was accomplished, I could perform a number of test flights over a range of three center-of-gravity positions: forward, center, and aft.

I based the aircraft at the Lancaster Municipal Airport (William J. Fox Field) the afternoon prior to flight and made a pre-dawn pre-flight inspection and launch. I sometimes took data just prior to sunrise in the smooth, cool air of the desert dawn. After a test flight, I landed at Edwards around mid-morning to drop off the data tape and load a new set of instrumentation batteries for the next morning. I often flew one or two jet flights during the day at Edwards and then returned the general aviation aircraft to Fox Field that afternoon in preparation for the next test flight. With this procedure, I accomplished a handling qualities evaluation program on a specific general aviation aircraft in about four months, which included instrumentation and data reduction. Fortunately, I had a test engineer assigned that flew with me most of the time to handle data reduction, which left me to my flying.

Paul Bikle advocated my flying the aircraft from the remote location at Fox Field. Paul took me aside and told me to get it out of the hangar and off on my own so I could get the work done much quicker. He was actually telling me to get out from under the test system that had been established at FRC. It was fine for more complex programs, but a hindrance for one of this size. I recall one sunny summer afternoon when I was on the FRC flight line preparing to fly one of my general aviation test aircraft back to Lancaster. One of our aircraft inspectors walked up to me and commented on the sandbags I had tied into the back of the aircraft for the next aft center-of-gravity test. The tiedowns did not meet his stringent specifications and he said that I could not fly the aircraft in that condition. As it happened, Bikle was on the ramp about 50 feet away, looking at another aircraft. Typically, Paul had an unlit cigar in his mouth as he overlooked his domain. I told the flight inspector, "If you get that short, fat guy with the cigar to say I can't fly, then O.K., otherwise, I'm on my way to Fox Airport." The inspector was aware of the general aviation program, and was already a little bent out of shape at being excluded. Rather than challenge the boss, however, he walked back into the hangar and I went on with my test program. Even though it was a small program, it was a great experience to be able to conduct it practically on my own.

The program provided basic stability and control data to verify that the airplanes were generally airworthy. However, as history has proved over the years, aircraft often have characteristics that can prove a detriment in certain flight situations. If something can happen, it is just a matter of time until it does happen. On one aircraft, I discovered a rather unique characteristic in the landing approach configuration. A small stick-forward pulse on the yoke resulted in the elevator floating to the full forward stop. This caused a sudden nose-down condition and dive. I could usually recover with a moderate force on the yoke, restoring normal flight control. When I presented the aircraft manufacturers with this data, they admitted to knowing about this trait. They actually knew of an aircraft crash in this model aircraft during a night approach in a light rainstorm. The pilot, finding himself too high after breaking out of the clouds, pushed forward on the stick. He failed to recover and came to a stop in a private garage some distance short of the runway. Both the pilot and the passenger received minor injuries. That particular flight characteristic was not mentioned in the aircraft handbook.

General aviation manufacturers were reluctant to publicize anything negative about their aircraft for fear of losing sales. This aircraft had been in service for a number of years, and I'm sure if I could find one today, it would hold the same potential danger that it did years ago. Company test pilots sometimes told me that they always had a difficult time getting a warning statement into the aircraft operator's handbook because their sales department felt this would make it more difficult for them to compete with other manufacturers. "Don't scare off the buyer," seemed to be their philosophy, even if it meant withholding critical safety information. I hope this is no longer the case. The military and airlines have always been diligent in giving their pilots every available piece of information concerning the safe operation of the aircraft. One of the most significant contributions from our NASA general aviation program was to improve handbooks and operating manuals by including more handling qualities data and safety warnings.

7

The Best of Times, the Worst of Times

Introducing the Lunar Landing Research Vehicle

Without question, the Lunar Landing Research Vehicle (LLRV) was the most unusual machine I ever flew. It evolved naturally as part of the Apollo project. A mission to send men to the Moon and return them safely to Earth seemed nearly insurmountable with its many critical phases, not the least of which was the final landing on the lunar surface. Many mission planners anticipated that this phase would be automatic, with the astronauts just along for the ride. Ultimately, they decided to allow the astronauts to manually control the final landing maneuvers. In 1961, Hubert "Jake" Drake, chief of the NASA FRC Advanced Programs office, convinced Walt Williams (then Associate Director of the Manned Spacecraft Center) that a free-flight simulator was needed to evaluate the problems of a manned lunar landing. It would also serve to train the actual crews who would fly the missions. Subsequently, NASA contracted with Bell Aerosystems to build a flying simulator, the LLRV.

The LLRV had to simulate a vehicle's performance in the Moon's airless, low-gravity (only 1/6 that of Earth's) environment while flying here on earth. Initially, Bell Aerosystems built two LLRVs for NASA to test at Edwards. They arrived at NASA FRC in the spring of 1964. Joe Walker and I were assigned as the project pilots. We were joined later by Army Maj. Emil "Jack" Kluever. Joseph S. Algranti and Harold E. "Bud" Ream from the Manned Spacecraft Center flew the vehicles in preparation for their use as Lunar Landing Training Vehicles (LLTV) in Houston. Usually, when government procures aircraft from a civilian contractor, the initial test flights of the new vehicle are made by contractor test pilots. In this case, the vehicles were turned directly over to NASA for initial evaluation.

The LLRV was a most unusual vehicle. It looked like a 10-foot-high boxy mass of aluminum pipes welded together with four legs. It had a pilot's seat on the front end balanced by a platform, extending to the rear, that held flight computers. A General Electric CF-700-2V turbofan engine, which developed about 4,200 pounds of thrust, occupied the center of the vehicle. The engine was mounted on gimbals, allowing it to pivot in the pitch and roll axes. Also in the center of the LLRV, on either side of the engine, were two large peroxide-fueled lift rockets of 100-pounds to 500-pounds of thrust each. Smaller peroxide-fueled rockets provided attitude control. The thrust of these attitude rockets was not adjustable during flight. Technicians had to pre-set them on the ground. There were two sets of control rockets and the pilot could select

NASA photo EC64-00580

Mallick contemplates the Lunar Landing Research Vehicle (LLRV) prior to a flight. It resembled a plumber's nightmare on shopping cart wheels.

one or both for use during flight. This allowed flight evaluation of various control power levels, one of the prime areas of interest. Attitude rockets, operated in pairs, were located on each of the four corners of the LLRV frame to provide adequate moments for roll, pitch, and yaw control. Just outboard of the engine were two spherical peroxide tanks that provided fuel for the lift and the attitude-control rockets. A JP-4 kerosene tank provided fuel for the jet engine. Two smaller spherical tanks to the rear were high-pressure helium containers to pressurize the peroxide-fuel tanks. The LLRV was an odd looking contraption indeed. It didn't look like it should fly, but it did fly. It wasn't too bad, once I had the hang of it. A digital computer on the rear platform kept everything in order and provided input to a fly-by-wire control system.

 A typical test flight in the LLRV was a complex process. Ground crews powered up the LLRV's electrical system several hours before the pilot entered the cockpit. They then completed a set of electronic systems checks and allowed the computers to function for several hours to provide confidence that they were stabilized and ready for flight. The pilot then entered the cockpit and began his pre-flight checks. He started the jet engine, checked the attitude control rockets, and verified that his control panel instruments were functional. Ground personnel in the control room also monitored the vehicle's systems. Instrumentation provided extensive data, some of which was not available to the pilot on his cockpit gauges. Therefore, the ground monitors could sometimes detect problems before the pilot was aware of them. It was like having several co-pilots backing up the pilot. On a number of occasions, ground controllers picked up systems deviations that resulted in early termination of a flight. After the control room staff and the pilot were satisfied that the machine was operating properly, the pilot increased power and lifted off.

The digital fly-by-wire flight control system allowed the pilot to operate the LLRV in two primary flight control modes: vertical takeoff and landing (VTOL) and Lunar Simulation Mode (LSM). In VTOL mode, the pilot locked the engine gimbals and the turbofan engine provided all the lift. In LSM, the gimbals were unlocked and the pilot throttled the jet engine specifically to lift 5/6 of the vehicle's weight, simulating lunar gravity conditions. The computer continuously adjusted the thrust as jet and peroxide fuel burned off. In each case, the peroxide thrusters provided attitude control.

Computer-controlled engine angle in LSM allowed the pilot to maneuver the LLRV while adjusting the lift rocket engines to fly in the simulated lunar gravity field to a designated landing spot. Sensors measured and compensated for atmospheric drag on the vehicle. The vehicle had some undesired aerodynamic pitching moments as a result of its speed through the air. The computer damping system increased the firing of the attitude rockets to cancel any angular rate that these aerodynamic moments generated. All in all, the LLRV provided a very good simulation of the final approach and touchdown in a lunar environment.

The LLRV control panel included a conventional center stick that provided pitch and roll control, and conventional rudder pedals for yaw control. A conventional jet engine throttle was located to the pilot's left as in a fighter plane. A stick, similar to collective control in a helicopter, actuated the lift rockets. It was attached to the floor on the pilot's left side.

To lift off in the VTOL mode, the pilot held directional control with the rudders and advanced the jet throttle until the LLRV rose on its leg struts. As the LLRV cleared the ground, he used the peroxide attitude rockets to provide pitch and roll control. He then tilted the LLRV forward like a helicopter and climbed to about 1,500 feet above the ground and approximately 2,500 to 3,000 feet from the touchdown point to begin the lunar simulation. Average flight speed was around 50 miles per hour. Speeds much above 60 miles per hour resulted in the pitch attitude rockets firing almost full-time to counter the aerodynamic pitching moment. This left very little pitch control for damping or pilot control inputs. Since the pilot had no direct way of knowing precisely how often the attitude rockets fired, he relied on the ground monitors to alert him by radio when he was approaching 80 percent of the pitch control capability. This determined the maximum allowable airspeed.

NASA photo EC64-00541

A powerful turbofan engine propelled the LLRV to hover altitude. The pilot could then practice simulated landing approaches, using reaction control thrusters to maneuver.

Up to this point, flying of the LLRV was very much like flying a helicopter, but with greater limits on airspeed. The LLRV was responsive and quite controllable within these limits in the VTOL mode. Once the pilot stabilized the LLRV at 1,500 feet above the ground and around 50 miles per hour, he selected the Lunar Simulation Mode and fired the lift rockets. The computer automatically weighed the LLRV at this point and reduced the jet engine throttle until the total thrust compensated for 5/6 of the vehicle's weight. At the same time, it unlocked the engine gimbals and controlled the thrust angle of the jet engine to allow the LLRV frame and lift rockets to maneuver without causing unwanted linear translations (horizontal movements).

Flying this ungainly vehicle took some getting used to. I found it an interesting challenge to fly, hover, and land on a desired spot. Like all tasks, the more I did it, the better I became. Controlling a vertical descent with the lift rockets was exciting too, as it seemed to take a long time to arrest the vehicle's vertical motion as I approached the ground. Flying the LLRV proved that an astronaut could operate a vehicle in the airless lunar environment with 1/6 Earth's gravity, but it required a little adaptation by the pilot to this new realm of flight.

One of our primary goals with the LLRV was to evaluate control power settings. We needed to know just how little control moment we could get away with and still safely operate the vehicle. Technicians decreased the thrust on one set of attitude control rockets prior to each flight as we sought the minimum acceptable power level. Joe and I then flew the standard lunar landing profile and evaluated the acceptability of each control power setting. We always had the second set of attitude controls adjusted to a control power setting known to provide safe flight conditions. If we entered a dangerous control situation, we rotated a switch, located on the left-hand side panel, to restore a more powerful attitude control and allow us to safely recover the LLRV. Joe and I took turns flying with various control power settings. After each flight, we made qualitative pilot evaluations using the Cooper-Harper rating scale, a standard flying qualities rating system. Ground personnel and engineers also used extensive recorded data, including pilot inputs and vehicle response, to verify our evaluation of a particular control setting.

Flying with the Boss

At this time, Joe Walker was the chief pilot at FRC and my boss. The LLRV was the first program that I shared with him as a co-project pilot. Unfortunately, it was also the only program. Joe, being the senior pilot, flew the first flight on the LLRV on 30 October 1964. After he had flown several times, I made my first flight on 9 December. I felt honored to participate in the LLRV program and enjoyed flying with Joe. He was a world-renowned research pilot, recognized and honored by the aeronautical community. President John F. Kennedy had presented him with the 1961 Collier Trophy for his work with

NASA photo E65-12573

Mallick (left) shared LLRV flying duty with then Flight Research Center Chief Pilot Joe Walker. An accomplished research pilot, Walker was highly respected in the flight test community.

the X-15. Yet, in spite of all of his accomplishments and the recognition he had received, Joe was a real "down-to-earth" person and one with whom I got along very well. His behavior was always very professional and he expected the same of his pilots. Joe's management style could best be described as: "Lead by example."

Joe and I attended nearly all of the LLRV planning meetings prior to the first flight. We were involved in cockpit design, instrument layout, and all the developmental details. The LLRV had a lightweight ejection seat, designed and built by the Weber Company in Los Angeles. An excellent design, it was even better than a zero-zero seat. A zero-zero ejection system is one that will eject the pilot safely from an aircraft sitting on the ground at zero speed and zero altitude. The Weber seat could get the pilot clear of the vehicle very close to the ground, even with a 400-foot-per-minute rate of descent. To accomplish this, the center of gravity of the pilot/seat combination was critical. Each pilot was fitted with a custom seat cushion that was of a height that complimented his body weight and a density that allowed the seat rocket motor to propel the pilot high enough above the ground for safe parachute deployment.

To properly fit the special seat cushion, each pilot was placed in a kind of swing device. Here, his individual center-of-gravity (c.g) was determined by taking readings at various angles of the swing. Depending on the pilot's weight and mass distribution, his personal c.g. was centered at a unique location, usually somewhere deep in the stomach behind his navel. The measurements were rather time-consuming and, as Joe and I each took our turn in the swing, we noticed that an unusual number of very attractive ladies were passing the doors to the laboratory that we were in. Most turned and smiled as they went by. After the tests were finished, Joe and I confessed to the Weber engineers that we hadn't minded the long time required for the tests because we enjoyed the "girl watching" opportunities. We also commented on their savvy in locating the laboratory in such a strategic spot for such observations. "Fellows," they said with a laugh, "you weren't the ones watching. Those gals heard we had two test pilots in for testing and they came down here to get a look at you. We hardly ever see them on a normal workday." I think that was the first time I was a victim of "boy watching." I don't know about Joe.

Test flying can be a hazardous profession. It doesn't always appear so because numerous people work hard to eliminate as many risks as possible and manage the rest. It is impossible to eliminate every risk and still fly the test aircraft, so the test team accepts a certain amount of "calculated risk." Day to day activities of the operation can take on the appearance of routine, but then something completely unexpected jumps out of the woodwork to remind everyone that it can be a risky business. Probably every test pilot has had an experience or two with a new aircraft or system that gave him a good scare, or at least got his attention.

My most memorable experience in the LLRV occurred during an approach with limited attitude control power. As I arrested my forward motion into a hover and added more collective (lift stick), the LLRV entered an uncontrolled roll to the left about 30 feet above the ground. I immediately scrambled for the attitude selector control and activated both attitude control systems. The response was immediate, but not what I had hoped for. Now, just 20 feet above the ground, I suddenly entered a high-rate roll to the right. I had overcontrolled and, after several swings back and forth, I damped out the oscillation and safely landed the vehicle. This one really "watered my eyes." I had very nearly crashed! The engineers were as stunned as I was. They immediately started searching for the cause of the initial left roll.

A look at the recorded data revealed that I had inadvertently pressed down on a peroxide trim switch located on the end of the collective. I had apparently depressed the switch shortly after entering the Lunar Simulation Mode. The switch was designed to control a lateral trim problem between the peroxide-fuel tanks mounted left and right of the vehicle's centerline. Although part of the original design, flight test had revealed that the switch was not really needed and it was seldom used. Since the fuel burn was symmetrical, trimming was not required. This incident occurred after I had logged about 25 flights and I was at a loss as to why I had accidentally hit the switch. After some thought and "soul searching," I came up with an explanation.

Just prior to this test flight, I had been issued another ejection seat cushion from the Weber Company. Apparently their follow-on testing indicated that I needed a thicker cushion. The new cushion was about two inches thicker and raised my body position relative to the collective stick. Where my left hand had once rested several inches from the top of the collective, it was now very near the top. My clumsy old thumb curled right over the end of the collective where the lateral trim switch was located. The trim switch had a very-light-touch, spring-loaded center position. During the flight, my thumb pressed lightly on the switch, driving the LLRV into a lateral out-of-trim condition. This knowledge restored my confidence and, from that flight on, I kept my thumb away from the trim switch. Technicians also increased spring force in the switch to prevent inadvertent actuation. It is always interesting, after the fact, to discover how something so simple could lead to such a dangerous situation.

One of the hallmarks of the NASA flight research was sharing information. If a pilot encountered some difficulty, it was analyzed with the idea that someone else could benefit from the experience and perhaps avoid a problem in the future. Joe Walker had an experience with the LLRV that provided me with useful information that helped me later on.

After we had flown the LLRV for about a year, FRC management asked Joe to fly a lunar simulation profile for a demonstration. This request, for what pilots referred to as "show and tell," was not unusual for a "mature" research program. We often put on a demonstration flight for visiting Headquarters directors and managers or other VIPs. I was never really fond of these demonstration flights, but I knew it was necessary for political reasons. Over the years in my testing career, I had personally observed several test pilots killed in the process of demonstrating their machines. Some were mechanical failures that could have happened at any time, and some resulted from over extension of the test pilot's skill while trying to make their vehicle seem more impressive.

We had conducted all of our research flights on the South Base area for convenience and safety. For the demo, Joe was to fly the LLRV over the NASA taxiway near our facility at the northern part of the base. The LLRV was to come into a hover and land in the NASA aircraft parking area not far from the main building and hangars. Joe entered Lunar Simulation Mode and started his approach to the landing spot. It was obvious to me as a ground observer, and knowledgeable LLRV pilot, that he was going to overshoot the landing site. At our normal test area on South Base, this was not a problem, as adequate space was available for the overshoot. In this case, there was very little excess room on the planned flight profile. Joe realized his problem in plenty of time, and switched the LLRV out of the lunar simulation mode into the VTOL mode, which enabled the jet engine to arrest the horizontal translation. He brought the LLRV to an uneventful hover and landed on the planned spot or just slightly beyond it. A short time later, after Joe had left the LLRV program, I was asked to fly a similar demonstration. I did, but I planned the touchdown point about 100 yards further out from the hangars and people to provide a greater safety buffer zone. The demonstration went well. We had learned from Joe's experience.

I had a lot of early "reveilles" in 1965 for LLRV flights as well as for a new son, Darren, born in March. Audrey and I decided that he would be the caboose on the "Mallick train." We had purchased our new home in California in September 1963 and we had ample room, but the thought of raising more than four children and facing the costs of college was a little sobering. Four children seemed like a nice size for a family, and so it was.

From October 1964 until late 1965, Joe Walker and I flew the LLRV, conducting data verification flights along with advanced research flights to evaluate minimum and safe control power settings for a lunar landing. Toward the end of 1965, an Army pilot named Emil "Jack" Kleuver joined the NASA Dryden Pilot's office and the LLRV test program. About the same time, Joe Walker left the LLRV program and began preparations to fly the XB-70. NASA planned to use one of the U.S. Air Force XB-70 aircraft to conduct high-altitude, high-speed research. Two of the large jet bomber prototypes were built, but the design was not put into production. NASA saw the aircraft as a good platform with which to study supersonic transport technology. I continued to fly the LLRV into the spring of 1966, at which time the research portion of the flying was nearly complete. Plans were underway to transfer the two LLRVs to the NASA Manned Spacecraft Center in Houston, Texas. There, they would serve the Apollo astronauts as Lunar Landing Training Vehicles (LLTV). Two Houston pilots, Joe Algranti and Harold "Bud" Ream, came out to NASA FRC to be checked out as instructor pilots on the LLTVs. I had flown with both men, some years back, when we were all at NASA Langley.

In the spring of 1966, I phased out of the LLRV program. Before I did, however, I had the opportunity to fly the first flight with the three-axis side-stick controller on LLRV number one in March 1966. The controller was similar to that which was to be used in the actual Lunar Excursion Module that would land on the Moon. Adapting to the three-axis controller was not difficult. It was a side-stick with the normal pitch and roll inputs, as well as a twisting motion around the vertical axis of the controller to provide yaw inputs. Jack Kleuver made a number of research flights to study the visibility restrictions that the astronauts would face. He also helped with the checkout of Algranti and Ream.

Simulators have always been important to flight research. Initially, simulator's served as procedure trainers, but they gradually evolved. As they

The Best of Times, the Worst of Times

grew more complex and sophisticated, simulators came to more accurately reflect the aircraft they simulated. During the period when I flew the LLRV, I had the opportunity to visit NASA Langley and fly their Lunar Landing Simulator. The Langley simulator was not a free-flight vehicle, but one suspended by cables in a large gantry. A crane mechanism provided the translational motion for the lander in response to pilot inputs through a computer interface. I was generally impressed with the similarity to LLRV handling qualities. My only complaint was that an unusual cockpit motion, caused by cable swing and the travel limitations within the gantry structure, made the Langley simulation fall short of the LLRV's performance. There was one other important difference. If I made a mistake in the Langley simulator, it dropped into a lock-up or reset mode, and the lander swung harmlessly on its cables. In the LLRV, there were no cables. A mistake could mean a fatal crash.

Despite that risk, I felt that the LLRV was as good a simulator as we could provide at the time. While it would give the astronauts a good feel for flight in the lunar environment, I had some misgivings about using it as a training vehicle at the time. Although we had learned a great deal, it was still a new and unique machine with about two years of flight experience. A number of key personnel from NASA FRC went to Houston with the LLRVs (now LLTVs), but they did not represent the complete body of experience that had been developed during our flight program. The aircraft went from a research environment, with all available backups and safeguards, to one of operational training with a great deal of schedule pressure. To provide more opportunities for astronaut training, the Manned Spacecraft Center procured three more LLTVs, slightly larger vehicles than the original LLRV. Regrettably, during the course of the training program, they lost two of the new vehicles and one of the originals to accidents. Fortunately, the excellent Weber ejection seats prevented any fatalities. The training proved useful, however, and the Apollo astronauts gained valuable flight experience in the LLTVs prior to their lunar missions.

I felt badly when I heard about the LLTV losses. It seemed as if we had passed something on that was unsafe, but this was really not the case. Everyone involved recognized that the LLRV/LLTV was a demanding vehicle to fly. The need for this training was so great that the risks of flying the LLTVs were justified. Neil Armstrong and Buzz Aldrin, the first astronauts to land on the Moon, proved this. Neil commented after the mission that he was very glad that he had the opportunity to fly the LLTV prior to his Moon landing. He had, in fact, made what was called a "manual mode" landing in order to avoid some rough lunar terrain that the automatic landing system had selected as a touchdown point. I thank Bell Aerosystems for building the LLRV, Weber for installing a fine ejection seat, and the good Lord for giving me the opportunity to be part of that program. I did not feel that I was "sitting in Neil's lap" when he landed, but having flown the LLRV during the initial development gave me the pride of participating in the overall lunar mission.

XB-70 Midair Collision, June 1966

After Joe Walker left the LLRV program and began preparing to fly the XB-70, I saw very little of him. I missed working with him, as we had become good friends as well as co-workers. Joe and I did not socialize with each other's families, but we did hunt together and we enjoyed an occasional cool beer together on a Friday afternoon. I think the fact that Joe and I both hailed from western Pennsylvania gave us a little common ground.

On the morning of 8 June 1966, I was scheduled for a weather check flight in an F-104. I always enjoyed flying these missions, especially the early morning flights when air traffic was light and I seemed to have the sky to myself. It usually meant getting up well before dawn in Lancaster, driving to Edwards, suiting up, checking the weather station, and taking off at or just before sunrise. There was no more accurate weather report than that from a pair of eyes that could survey the weather scene from the surface to 50,000 feet. A pilot in the high-performance F-104 could easily do that. Analyzing the weather was not always easy unless it was clear with unlimited visibility, or if it was completely unsatisfactory with no possibility of launching the mission. On this particular morning, it was in between. It was not quite bad enough to cancel the mission until another day, but not good enough to clear the mothership for takeoff either.

We had scheduled missions for the X-15 and the M2-F1, but the flight planners decided to put a weather hold on everything. I scheduled another F-104 flight to check the weather in about another hour or so. I was flying NASA 811, an F-104N, on my first mission and, as I called in the weather and the hold, I asked for another F-104N (NASA 812) to be readied for me in order to make a quick turn-around. As nice as the F-104s were, their fuel endurance was not great for this sort of mission. After I landed and parked my aircraft, I walked past Joe, who was getting in another F-104N, NASA 813. I had noticed his flight on the schedule board, but there was no indication of his planned mission. Joe waved and gave me his trademark big grin. Secretaries often commented that Joe was grumpy when faced with desk work. They were happy to see him get out of the office and fly because it improved his mood so much. I could understand this more in later years when I picked up the same responsibilities as chief pilot. In any event, I returned his greeting with a wave and smile. I then headed upstairs to give my detailed weather briefing and prepare to take out the second flight.

The maintenance crew prepared NASA 812 and I took off within the hour, returning to the planned X-15 launch area. Around Edwards, skies were blue with scattered clouds, but it was not so good in the launch area. I spent some time in the vicinity, but conditions continued to deteriorate. Noting my diminishing fuel supply, I recommended scrubbing the test flight and then I returned to Edwards. After I taxied to the parking apron and filled out the regular post-flight paperwork, I noticed two large black columns of smoke in

NASA photo EC77-08296

Mallick prepares for takeoff in the Bell Model 47G helicopter. Originally used to support the LLRV program, Mallick employed the Bell 47 to search for clues during the XB-70/F-104 accident investigation.

the desert east of Edwards. "That sure looks bad," I thought as I walked toward the hangar. I thought that it looked like someone had "augered in," which is to say, crashed. I dropped off my chute and helmet in the locker room and hurried up the stairs to the pilot's office. It was immediately obvious from the expressions of the people there that something really terrible had happened. Joe Vensel, the chief of operations, asked me if our helicopter, a Bell Model 47G was in flying status. It was. He directed me to take our flight surgeon, Dr. Jim Roman, out to the Camp Irwin area to a crash site that included the number two XB-70 and the F-104N that Joe Walker was flying. There had been a mid-air collision between the two and they had impacted in the hills, north of Barstow.

I was completely dumbfounded by this news. How could this have happened? I think I might have gone into some sort of mild shock if Vensel had not assigned me the helicopter mission. We used the Bell 47 to support the LLRV operation, and I was the only pilot available qualified to fly a helicopter at FRC. "Doc" Roman showed up quickly, and we boarded the "helo" just as the ground crewman topped it with fuel. Doc Roman and I were silent as we skimmed over the sage on the way to the crash site. We listened intently to radio chatter from Air Force rescue crews. I couldn't imagine how the crash occurred or why the pilots had not seen each other. I thought it was a classic mid-air collision, when two unrelated flights run into each other. I learned later that a special formation flight had been scheduled by the Air Force, at the request of the General Electric Company. Photographers had boarded a Learjet to take air-to-air photographs of the XB-70 flying in formation with a Navy F-4B, Air Force T-38A, Northrop YF-5A, and the NASA F-104N. All of these aircraft were equipped with General Electric engines, and the company planned to use the photos for a publicity campaign. Joe had been flying chase during a routine test mission. After the XB-70 had completed all scheduled test points, the two aircraft rendezvoused with the others for the photo formation.

As we approached the impact area, a thousand questions raced through my mind. Doc Roman seemed lost in his own thoughts. I think we were both overwhelmed by shock and disbelief, and only remained functional because we had a mission. I checked in on the Air Force "crash net" frequency and within a few minutes, we spotted the remains of the XB-70. The once proud ship was lying on the ground as if it "pancaked" in a flat spin. It was flattened into the desert, its formerly white surface charred gray and black. The majority of the fuel had burned off and the big black smoke columns were gone, replaced by thin streamers of white smoke and steam rising from what was left of the fuselage. The stainless steel airframe was flattened and shattered. The six round engine nozzles were now oval-shaped, but it was surprising how much of the structure had survived the fire.

We landed in a small clearing near some emergency vehicles and walked over to talk with the Air Force personnel. They indicated that co-pilot Maj. Carl Cross was still inside the XB-70 and had perished. North American Aviation test pilot Al White had managed to bail out, using his escape capsule, and was now on his way to the Edwards hospital. They could only point in the general direction of where they thought the F-104 had impacted.

The terrain in that area was like the rolling waves of an ocean, with swells and valleys. In some areas, the ridges were close together, making a helicopter landing in the wash difficult due to lack of rotor clearance. Doc Roman and I searched for about five minutes before we spotted the smoking fuselage of the F-104 laying in one of the valleys two miles northwest of the XB-70. We found a landing spot and hiked over to the wreckage. The F-104 fuselage was even flatter than the XB-70, and the center of it glowed white hot like burning magnesium. Doc Roman and I studied it for several minutes, trying to determine where the cockpit should have been. We decided that it was not there. We were looking at what was left of the fuselage from just behind the cockpit to the tail. The front section and wingtip fuel tanks were missing, as was the vertical and horizontal tail assembly.

We got back in the helicopter and resumed our search. After a few minutes, we found the cockpit section of the F-104, minus the radome. The fuselage had separated just behind the ejection seat. The cockpit section was lying on its left side near the top of a hill and Joe's body was still strapped in the seat. His helmet was missing and he was obviously dead. There was evidence of a fire, but it was out now. It appeared that the aircraft had broken up at altitude, scattering pieces over a wide area. Joe's injuries indicated that he had been killed instantly at the time of collision. Doc Roman completed his examination and we returned to the helicopter to call the Air Force rescue personnel at the XB-70 site. We reported our findings and the Air Force dispatched a team to recover Joe's body.

About this time, the crash site coordinator notified us of an unauthorized civilian helicopter approaching the area. We acknowledged this and lifted off with the intention of flying to nearby Barstow airport to refuel before returning to Edwards. As I lifted into the air, I spotted the unauthorized helicopter, a Bell Jet Ranger, flying directly toward the F-104 cockpit impact site. I was in a very emotional state as I had just seen a very good friend of mine who had been killed in a terrible crash. I figured this might be a news helicopter filled with photographers and I did not think it appropriate that they should enter a restricted crash area and take photographs.

I took up a heading directly toward the Jet Ranger just as teenagers would do with automobiles in a game of "chicken." I wanted this guy out of there, so I held my course with grim determination. The other pilot veered away at the very last moment, and I think we may have passed within 20 feet of each other. With his speed advantage, he flew away from us and turned to come in again on another course. I cut across the circle and again lined up between the F-104 cockpit wreckage and the Jet Ranger. Once again, I held my course until he broke away. After this pass, I think that he got the message. He was not going to get any pictures this day. Shortly afterward, he turned and headed out of the area. I thought that perhaps Doc Roman would be concerned and disapprove of my actions in this head-on challenge, but he only smiled at me. I think that he would have done the same if he had been flying.

I reported the encounter to the Air Force crash site coordinator and headed to Barstow for fuel. As we approached Barstow for landing, I spotted a Jet Ranger on the ramp that had just landed. After I shut down, I considered looking up the pilot, and breaking his neck. Suddenly, I realized that I was completely drained, emotionally, and extremely tired. Everything that had happened was finally sinking in, and I knew that the NASA officials were waiting for our return to Edwards. I took the time to write down the Jet Ranger's identification number and, the next day when things slowed up a little, I passed the number on to the Air Force and the FAA. One week later, the FAA called me and told me that they had traced the aircraft to Van Nuys and found the pilot. His flying license was suspended for one year. I was happy that some action was taken. I

Courtesy of the Air Force Flight Test Center History Office

Following a research mission, the XB-70 formed up with several other aircraft to provide a publicity photo opportunity for the General Electric Company. Joe Walker's NASA F-104N is positioned below the bomber's right wingtip.

didn't want him to get away with ignoring the rules in order to make a profit from someone else's tragedy.

XB-70 Accident Investigation

The next weeks and months were hectic and stressful for all of those associated with the accident. Losing two aircraft and two lives was almost unthinkable, but it had happened and now we had to deal with it. The Air Force and NASA formed an accident investigation board. Don Bellman represented NASA as a non-voting member. Don was an excellent engineer and manager who had been project manager for the LLRV from its first days. Although Don was the only official NASA member on the board, a number of other NASA specialists, including me, assisted him in gathering evidence and supporting documents. As a member of the Flight Operations office, I searched through Joe Walker's flight records and F-104 maintenance information. The majority of my time in the first few weeks involved flying the helicopter to the crash site and looking for physical evidence that might explain why the accident occurred. Don Bellman put the pieces together and then came back to me with new instructions. "Don," he would say, "I really need the upper piece of the left horizontal stabilizer from the F-104." Then I would head back out to the

The Best of Times, the Worst of Times

Courtesy of the Air Force Materiel Command History Office

The XB-70's right wingtip hangs above Joe Walker's F-104N like the Sword of Damocles. An inadvertent control stick movement may have caused the F-104 to enter the powerful wingtip vortex.

area to find that piece. Locating the pieces was difficult as they consisted of small fragments, scattered over several square miles of desert.

The accident investigation board included eight voting members from the Air Force. The non-voting members included four officers, two NCOs, and six civilians. It seemed that the board's initial attitude was that this was likely a case of pilot error, and that Joe Walker would take the blame. Those of us who knew Joe were not ready to accept this without careful examination of the evidence. Some of our F-104 aircraft had experienced unexplained pitch transients in flight. These incidents were very rare and we could never duplicate them during ground tests. The board examined all maintenance records for the incident aircraft (NASA 813), and studied the remains of the control system that we recovered from the crash site. None of the evidence suggested that there had been any control problem.

The accident sequence took place suddenly. Because it occurred during a photo mission, there was substantial documentation for the board to examine. A motion picture camera had recorded all events until just before the collision,

Courtesy of the Air Force Flight Test Center History Office

Caught in the bomber's wingtip vortex, the F-104N rolled across the top of the XB-70, shearing off the bomber's tail fins.

and then picked up again shortly afterwards. There were also numerous still photos of the accident sequence. The images showed that the F-104 burst upward from its position, tearing through the XB-70's right wingtip. From there, the F-104 rolled inverted as it passed across the top of the XB-70, shearing off both of the ship's large vertical tails. By this point, the F-104 was in several large pieces and trailing a ball of fire from its ruptured fuel tanks.

General Electric chief test pilot John Fritz, flying on Joe's right wing in the YF-5A, saw nothing amiss prior to the collision. After the impact, the XB-70 continued in straight and level flight for 16 seconds as if nothing had happened. Someone got on the radio and shouted, "Mid-air, mid-air, mid-air!" Colonel Joe Cotton, riding in the back seat of Captain Pete Hoag's T-38, called frantically to the XB-70 crew. "You got the verticals came off, left and right. We're staying with you. No sweat. Now you're looking good." But, the giant ship rolled ponderously to the right and entered an inverted spiral. As it shed parts, fuel poured from its broken right wing. Cotton and the others began shouting for the stricken crew to eject. "Bail out. Bail out, bail out!"

After some difficulty, Al White ejected in his escape capsule. Maj. Carl Cross, possibly incapacitated by extreme g-forces, was less fortunate.

Additionally, a seat retraction mechanism failed, making automatic encapsulation impossible. He remained trapped in the stricken jet and died when it struck the ground. I did not know Carl personally, but I understand that he was a top-notch flyer and a fine gentleman. He had recently returned from a tour of duty in Vietnam and this was his first checkout flight in the XB-70. I thought of Carl as I approached my first flight in the XB-70 about a year later. I believe in God, but I wonder at times just what the master plan is. Why do so many young men have to die in military service? It has been going on almost since the beginning of time, but why? Carl had survived combat duty in Southeast Asia. He returned to "safe" peacetime test flying at home. This flight, even the photo formation, was a fairly routine operation. Why?

The Air Force accident board was nearly ready to write up their conclusions. They had all sorts of recorded and photographic data, and thousands of pages of documents. There was substantial political pressure to close the investigation. People wanted answers and they wanted them now. Don Bellman was not satisfied that we had all of the facts. He continued to put the pressure on me to find the upper surface of the left tip from the F-104's horizontal stabilizer. He had suspicions about the manner of contact between the two aircraft. I enlisted the aid of project engineer Jim Adkins to help search for the elusive piece of F-104 hiding somewhere out in the desert. We flew several flights a day in the helicopter, occasionally stopping at Barstow for fuel and then returning to the crash site to hunt some more. Jim and I felt that we had picked up almost every piece but the desired one. On the fourth day, we spotted a shiny piece of aluminum. Both Jim and I recognized it, even from the air, and we knew that we had finally found the missing piece of Don Bellman's puzzle.

Jim was a former P-51 pilot. After World War II, he had entered the field of engineering. He was familiar with aircraft and I had briefed him thoroughly regarding the helicopter and its unique safety hazards. He was so excited about spotting the missing piece that, as I landed in a valley, Jim jumped out and ran up the hillside. As he did, his head came within about 12 inches of the helicopter's spinning rotor blades. It happened so fast that it was all over in an instant. I sat there picturing another death associated with the XB-70, and I just thanked God that Jim had cleared the rotor safely. When he came back over the hill with the horizontal stabilizer part, I didn't have the heart to give him too much hell, but I did mention it. He just looked at me as if to say "Big deal, I didn't hit it did I?" Maybe life is like that. I guess some people just have more luck than others. Jim used up one of his "silver bullets" that day.

Don made a careful study of the stabilizer fragment. On the top outer edge was an imprint of the XB-70's right wingtip light. Don's theory was that the first part of the F-104 to contact the XB-70 was the top of the F-104's left horizontal stabilizer, which came up under the right wingtip of the XB-70. The F-104's horizontal tail helps keep the aircraft in straight-and-level flight and provides pitch control for maneuvering. Don requested data from North

The Smell of Kerosene

Courtesy of the Air Force Flight Test Center History Office

The smoke plume from the XB-70 was visible for many miles. Only pilot Al White survived the mid-air collision.

American Aviation and also made his own calculations to determine the energy of the wake vortex flow around the XB-70 wingtip. Don produced his results first, beating the engineers and the North American computer. Both results were similar and they indicated that within about eight feet of the XB-70 wingtip, the vortex energy was such that it would equal the F-104's full roll control at the same airspeed. Once the horizontal tail of the F-104 came up under the wingtip of the XB-70, it became "pinned" by the wingtip vortex from the bomber. The F-104 lost its trim and pitched up violently, rolling inverted across the top of the bomber. The accident board concluded that the swirling wake vortex only became a contributory factor in the accident after the F-104's tail was so close to the XB-70 that a collision was imminent.

We still needed to find out why the F-104 was close enough to touch the XB-70. The board postulated numerous theories. Joe had a reputation as a fine test pilot and a levelheaded professional, dedicated, and safety conscious. He had nearly 5,000 hours of flight experience and had flown chase for the XB-70 nine times, eight in an F-104. It was hard to imagine that lack of formation proficiency, or some lapse of judgement could have led to the disaster. In the end, the leading theory held that Joe had become distracted somehow and had caused the F-104 to move slightly, imperceptibly, toward the XB-70.

During the course of the flight, ground controllers had alerted the formation to the presence of other air traffic in the area. An Air Force F-104D, a two-place aircraft with a photographer in the back seat had received permission to join the formation for a few pictures before returning to Edwards after a separate mission. Don Sorlie, the pilot of the F-104D later reported that the formation looked good, although the two aircraft on the left were not flying as close as the F-104 and the YF-5A on the right. This indicated that Joe was flying a close wing position. Shortly before the collision, Edwards reported a B-58 approaching the formation. It would pass high above, in the supersonic corridor, and wouldn't pose a hazard. Several pilots in the formation responded that they could see the B-58's contrail. Joe never made such a call. He may have been attempting to spot the B-58 at the time his aircraft collided with the XB-70. The board ultimately concluded that Joe's position relative to the XB-70 left him with no good visual reference points for judging his distance. Therefore, a gradual movement in any direction would not have been noticeable to him. The board blamed an "inadvertent movement" of the F-104 that placed it in a position such that "contact was inevitable."

I had no argument with the findings, but as a test pilot, I always wanted to know the exact cause. Why had Joe allowed the "inadvertent" movement to happen? The length of the precision formation mission may have been a factor. Cloudy weather had extended the flight time and forced the formation to move to a different area than had originally been planned. Joe had been flying close to the bomber's wing, in a position that made it difficult to judge his distance. Other air traffic in the area created distractions. Accidents are usually the result of several factors, so we will never really know for sure what happened that day. Fate simply caught up with two fine aviators and shocked the flight test community.

Others suffered for their role in organizing, planning, and approving the flight. John Fritz, chief test pilot for General Electric, set the chain of events in motion when he requested permission for the photo mission from a North American Aviation representative. Subsequently, Col. Joe Cotton, XB-70 Test Director, agreed to provide such an opportunity on a non-interference basis following a regularly scheduled test flight. Another North American official disapproved the photo flight, citing a tight schedule, but Cotton and Fritz lobbied to include it on an upcoming flight. Both North American and Cotton's immediate supervisors, Col. Albert Cate and John McCollom, finally approved the photo opportunity. No further approval was sought from higher headquarters. Cotton arranged to include an Air Force T-38, a chase aircraft that normally accompanied the XB-70. Mr. Fritz requested a Navy F-4 from Point Mugu Naval Air Station. It was authorized as a routine training flight in support of what was assumed to be an approved Air Force mission. Fritz himself piloted the YF-5A, bailed to General Electric by the Air Force. Although his officially stated purpose for the flight was to "perform engine airstart evaluations," he never actually did this. He tried to arrange for a B-58 to join the formation, and also for supplementary Air Force photo coverage, but was unsuccessful. General Electric contracted with Clay Lacy for use of the Learjet photo plane. Cotton supported Fritz in a request for the NASA F-104. As chief pilot at NASA FRC, Joe Walker was within his authority to schedule the chase operation, but his superiors were not aware of the photographic mission.

An Air Force public affairs officer in Los Angeles learned of the photo mission just two days before the flight, through a call from a commercial photographer. He referred the caller to Col. James G. Smith, chief of public affairs at Edwards. Smith had also been unaware of the planned formation flight, but voiced no objections once he ascertained that Col. Cate had approved the mission. The accident investigation board later ruled that Col. Smith should have advised the responsible parties of proper procedures for approving such a mission through higher headquarters. A memorandum, dated 12 August 1966, from Air Force Secretary Harold Brown to the Secretary of Defense concluded:

"The photographic mission would not have occurred if Col. Cotton had refused the General Electric request or at least not caused North American to reconsider its reluctance. It would not have occurred if Col. Cate had taken a more limited view of his own approval authority. It would not have occurred if Col. Smith had advised of the need for higher approval. It would not have occurred if Mr. McCollom had exercised the power he personally possessed to stop the flight. But it did occur."

Secretary Brown further stated that "these individuals acted in ignorance of prescribed procedures, rather than with intent to violate them." He noted that the commander of Air Force Systems Command, with the concurrence of the Air Force Chief of Staff, directed a number of disciplinary actions against the responsible parties. Col. Cate was relieved as Deputy for Systems Test and reassigned. Col. Cotton and Col. Smith received written reprimands, as did John McCollom. The Air Force also made numerous administrative changes to improve operational procedures.

Joe Cotton later ended his career as a colonel, but I'm sure that he would have been promoted to general officer level had the accident not occurred. I knew Joe well, and flew the XB-70 with him once in 1968. I feel that the Air Force lost a valuable resource when Joe left the military.

Following the accident, individuals and organizations paid numerous tributes to Joe Walker and Carl Cross, citing their contributions to aviation. President Lyndon Johnson personally honored Joe for his work on the X-15 program. Joe's wife Grace filed a lawsuit against the General Electric and the government because of the loss of her husband and father of her children. Although I couldn't bring myself to say that I really didn't believe in doing something like that, I asked her about it. She said that any settlement would be subtracted from her government benefits, and she would not receive anything extra. She was just so upset about losing Joe and she felt that there were other people and organizations responsible for the accident. She also wanted to protect Joe's good name. Not everyone had showered praise on Joe. Some were laying blame. That didn't seem right. Hearing rumors about Joe's alleged "screw up" made my blood boil. Someone once passed a second-hand story about such a negative comment to me. I won't repeat the comment or identify the speaker, but I once mentioned it to a fellow pilot and he said to me, "Don, you must remember, even a cur can piss on a dead lion." I thought about that for a minute and it made me feel a little better.

Joe Walker was a fine aviator and I was privileged to fly with him on numerous occasions. Joe was also a good friend. I hunted with him, drank with him, and worked with him. I suffered greatly when he died. When I stood beside his body on that black day in the desert, my heart sank and a part of me died too. Later, I thought again of the death of Bill Alford from Langley, and others I knew who perished in aviation accidents. Those experiences use a person up, a little each time. It puts a wound in you that doesn't go away, even though you push it to the back of your mind and continue on.

Aftermath

In the aftermath of the accident, I helped revamp the Flight Operations office records and pilot's information. We discovered a lack of centralization in the records handling system. Qualification and flight currency data for NASA pilots were recorded in a number of places, but the information was not readily available. I worked up some pilot's dossiers similar to what we had used in the Navy. This made it easier to find the information whenever necessary.

Meanwhile, Center personnel battled the loss of morale that accompanies an accident. They bravely soldiered on as flight operations continued with the lifting body and X-15 programs. Work also continued with the LLRV, F-4A, JetStar, and numerous smaller projects.

Walker's death threw the FRC pilot's office into a state of turmoil. While the accident investigation board searched for answers, we struggled to find someone to fill Joe's shoes. Stan Butchart, one of our senior pilots was selected as the new chief pilot. He was respected and well liked, but only served in the position for a short time. In December 1966, Flight Operations Division director Joe Vensel retired. Again, personnel were shuffled around. Stan Butchart became Director of Flight Operations and Jack McKay replaced Stan as chief pilot. The office went through a lot of changes in a short time.

Joe Walker died while preparing for assignment to the joint USAF-NASA XB-70 program. I replaced Joe on this assignment, and Fitz Fulton, who was retiring from the Air Force to join NASA, became the primary XB-70 pilot. Fitz had flown the aircraft while in military service and was one of the top Air Force test pilots for bombers. I believe the only reason that Fitz retired from the Air Force was that he was due for assignment to senior officer's staff school, which was standard for career officers. Fitz loved to fly so much that he took the option of early retirement to join NASA as a research pilot. I feel that both Fitz and NASA benefited from that decision. He proved an excellent addition to the flight operations staff.

In September 1967, Jack McKay left the chief pilot's position for a flight safety assignment in the Director's office. Jack had suffered for some years from back injuries sustained in an X-15 accident in November 1962. Jack was a stocky Scotsman who didn't complain, but pain troubled him constantly and it became obvious that he could not maintain the day-to-day effort required of the chief pilot. Jack's emergency landing in the X-15 ended badly when the aircraft flipped over. The doctors found that Jack's spine was compressed, leaving him about an inch shorter than before the accident. He recovered and continued his flying career, but his condition deteriorated over time.

When Jack left the pilot's office, Center Director Paul Bikle approached me and asked if I would like to be considered for the position of chief pilot. My work, up to this time, was so interesting and time-consuming that I really hadn't considered moving to a position of greater responsibility. I had been at Edwards only four years, and there were other capable pilots who were qualified for the job. After thinking it over for a day, I told Paul that I would indeed like to be considered. I had decided that it was in the best interest for my career as a research pilot with NASA. The opportunity had presented itself much sooner than I had expected, but I felt I could handle it. When Paul selected me shortly thereafter, I was both pleased and surprised. I felt that my work following the XB-70 accident probably contributed to my selection. I had helped improve the Flight Operations record-keeping system and had worked closely with the

Air Force during the investigation. I believed my contributions made a difference during a challenging period in the history of the Center.

Some years later, I asked Mr. Bikle if any particular deciding factor caused him to select me. At that time, he really didn't know me as well as some of the other pilots under consideration. Paul smiled and said, "Don, I asked Chuck Yeager what he thought of you, and he said that you were a good, solid driver." Well, I hadn't really had any interaction with Chuck since my time at the Air Force Test Pilot School. Nevertheless, his opinion counted with Bikle. It always amazed me just how important something like this could be. Bikle had worked with Chuck at Edwards for years before joining NASA. He had a very high regard for Chuck as a test pilot and an officer. When Chuck called me a "good, solid driver," he was really saying that I was a highly qualified test pilot. It was high praise indeed, and resulted in my promotion. The world takes some unusual turns.

8

Flying the Heavies

Safety First

My greatest responsibilities as chief pilot were flight safety and the well-being of the pilots. All of them were highly skilled, and they conducted their research programs with a high degree of professionalism. I rarely had to question a pilot's actions. There were so many ongoing programs of a complicated nature that it was difficult to keep up with them all in detail. I compensated for this by letting the individual pilots know that they had to support me in this area. I directed the pilots to come to me with any program safety concerns, and I then reviewed the program and recommended changes I thought were appropriate. The pilots had a great deal of approval authority regarding flight safety matters, and nothing was flown or attempted if there was a question of safety. I let my pilots know that I wanted them to take a conservative approach to their programs. As we used to say in the Navy: "I didn't want any pilot busting his ass on my watch." The way I saw it, I had the watch all of the time.

Perhaps my experience with the Joe Walker accident had given me the attitude that never again did I want to go through something like that. I held the chief pilot position for 14 years before moving up in the organization. We conducted a great deal of productive and risky flight research during that time, and I never lost a pilot on my watch. I was proud of that fact, even though I realized that God and good luck were very much involved. Unfortunately, a few years after I left the chief pilot's position, we lost a NASA Dryden pilot in an aircraft crash. One is always too many, but the overall record was good.

Learning New Skills

During the XB-70 accident investigation, I never gave much thought to the future joint NASA-USAF XB-70 flight research program. Afterward, the focus shifted to the future. I was assigned as one of the NASA pilots for the remaining XB-70. Fitz Fulton was the primary NASA pilot and I was the NASA co-project pilot. Col. Joe Cotton was the primary Air Force pilot and Lt. Col. Emil "Ted" Sturmthal was the Air Force co-project pilot.

At the time I was assigned, I never had any misgivings about the XB-70 because no fault of the aircraft had contributed to the accident. It was a mixed feeling for me, however, coming into the program as a replacement for Joe Walker. Ship Two, the XB-70 that was destroyed, was the aircraft originally slated for the NASA-USAF Supersonic Transport (SST) research program. It had a number of improvements over Ship One, not the least of which was better construction and bonding of the stainless steel honeycomb sandwich aircraft skin. Ship One had a habit of losing wing surface skin when it cruised above Mach 2.6. Manufacturing techniques had improved prior to the construction of the second airframe. Ship Two also had an improved wing configuration, a better engine inlet ramp control system, and improved instrumentation. All of this was now academic. We inherited Ship One, the only remaining XB-70.

For many years, NASA FRC had been involved with small, high-speed aircraft. None of the pilots in the Flight Operations office had any prior experience in large aircraft. Consequently, NASA hired Fitz Fulton for his XB-70 experience. I needed to get up to speed quickly in order to contribute to the program. One of the more desirable features of my assignment was the opportunity to checkout in the XB-70 as first pilot and not merely function as a co-pilot. This made a great deal of difference in my attitude, and I was ready and willing to take all of the preparatory training available in order to achieve this goal. It was quite an opportunity for an ex-Navy fighter jock who had never flown anything bigger than the R5D or C-54. My main advantage was plenty of experience flying in two-pilot aircraft, and I was very familiar with the cockpit coordination that required.

The cost of operating the aircraft, over a million dollars per flight, made it hard to justify missions strictly for pilot checkout. Fortunately, there was another aircraft available that had similar flying characteristics: The Convair B-58 Hustler. I soon found myself assigned to an Air Force B-58 ground school in Little Rock Arkansas. The Air Force Flight Test Center at Edwards had a TB-58A (serial no. 55-0662) assigned in support of the XB-70 program. It was used for chase support and currency training for the XB-70 pilots. After I finished the B-58 ground school, Joe Cotton checked me out in the TB-58A. That was an experience in itself. The B-58 was a delta-wing, supersonic, medium bomber capable of carrying nuclear weapons. It had four afterburning J79 turbojet engines that enabled it to reach speeds in the Mach 2 range. The flight control system had a unique mixer assembly to provide proper control inputs throughout the envelope. Part of the preflight checklist included visually inspecting a compartment in the bottom rear of the aircraft that contained all of the mixer rods, cables, pulleys, and connectors that converted the pilot stick movements to the proper inputs for roll and pitch control. It was a sophisticated mechanical arrangement.

On one occasion, as Fitz and I were preflighting the aircraft, we arrived at the mixer. "Fitz," I asked him, "do you really understand what you are seeing up in the compartment and could you determine if anything was wrong just from the visual inspection?" Fitz just laughed and said, "Look for anything loose or hanging. That's about all you can do." I felt a little better because I could not do much more than that. I'm not sure the original designer could tell

Courtesy of the Air Force Flight Test Center History Office

Joe Cotton checked Mallick out in the Convair TB-58A Hustler. The delta-winged bomber served as a training surrogate (as well as a chase) plane for the XB-70.

with just a look. The area was so confusing and complicated that it was referred to as the "bicycle wreck." It looked as if about three bicycles had run into each other at high speed.

The checkout process on a new aircraft, even for an experienced pilot, is quite extensive. I had to learn all about the aircraft's design. I studied each of the systems: fuel, electrical, hydraulic, environmental, and flight controls. In the case of the B-58, I spent three weeks in classroom and flight simulator training prior to making a flight. The simulator was the best tool for learning normal and emergency procedures for each system. I also had to memorize all of the "boldface" sections of the flight manual. These parts, highlighted in boldface type, provided critical procedures for immediate response to inflight emergencies.

The B-58 had good flying qualities, and the men from Convair that designed the flight control system (including the bicycle wreck) knew what they were doing. The airplane handled well at high-speed and low-speed conditions. In the landing pattern, it handled just like a big fighter. I had the opportunity to

Flying the Heavies

fly the delta-winged F-106 at the Air Force Test Pilot School and it flew nicely. The TB-58A handled like the F-106 in the landing pattern. There were no flaps on either aircraft and the landing attitude in both was nose-high. Both the F-106 and TB-58A had a strong ground effect and it was easy to make smooth, precise landings.

The B-58 checkout was only the beginning of my road to the XB-70. It gave me experience operating a relatively large delta-wing aircraft, but it was still small compared to the XB-70. My next checkout was in a B-52D at Castle Air Force Base in northern California. The B-52 was not a delta-wing aircraft, but it was certainly large. The "commanders course" lasted a full month and included ground school, simulators, and about 35 hours of flight training. I had the opportunity to experience both night and day aerial refueling, even though neither was applicable to the XB-70.

Unlike the B-58, the B-52 did not resemble a fighter plane in any way. It was a big airplane, and challenging to fly. I had to learn a number of new techniques to handle it. Because the B-52 had large moments of inertia, both in the roll and pitch axes, it took some time for a control input to move the wings. On takeoff, the pilot was required to make a heavy back-yoke input for the initial rotation. Almost immediately after that, and on initial nose rotation, a forward-yoke input was necessary to prevent over-rotating the nose. These things resulted from the size and mass of the aircraft, and were really not that difficult to learn. After a short time, they seemed natural, in fact. The air-to-air refueling was another new experience for me. The slow roll response of the B-52 was more of a problem than the pitch response during refueling operations. To deal with this condition, we used a wing spoiler to provide a quicker roll response. It helped a great deal and reduced the pilot workload. Aerial refueling involved a very intense flying situation and I marveled at the Air Force instructors that made it look so easy. Maybe it would have been for me, too, after a few thousand more hours.

Ground handling of the B-52 was another new world for me. I had never taxied or maneuvered an aircraft this large on the ground. The long wingspan made it a special challenge. The wings had outrigger landing gear near the tips, and it was very difficult to estimate just where those outriggers were relative to taxiway lights and other obstructions. Braking the aircraft at low taxi speeds was also a task. Very gentle braking was necessary to prevent brake lock or chatter. Because the aircraft was so large, I found it difficult to estimate just how much brake I needed. I experienced better brake feel at higher ground speeds, and anti-skid brakes prevented wheel lock-up. All in all, the B-52 was really a big aircraft, and if that was the experience I needed to fly the XB-70, I was certainly getting it. Although my initial B-52 checkout was directly related to flying the XB-70, it paid off in later years for both NASA and me. As NASA took over the flight operation responsibility for the NB-52B mothership, I was assigned to participate as one of the pilots.

In the B-52, I learned to establish the proper glide path and directional alignment early in approach. With the large wingspan and inertia of the aircraft, I didn't want to have to make large corrections just before touchdown. This was much different from flying fighters. Smaller, more maneuverable aircraft could be easily corrected closer to the touchdown point. I had to learn a new set of skills for dealing with a large aircraft like the B-52. This helped me greatly, later on, when I flew the XB-70.

Flying the XB-70

Even after completing all this preparation and training, the XB-70 still seemed somewhat imposing to me. The aircraft was over 190 feet long. The wing span was 105 feet and it weighed over half a million pounds! The cockpit was so high in the air that we needed a two-story stairway to get onboard. The pilot's eye level was over 20 feet above the ground during taxi. The XB-70 had six large afterburning turbojet engines installed in two bays. A large air intake for each bay had a variable throat and bypass doors similar to a wind tunnel. A technician could walk nearly upright inside the throat. The inlets lowered the speed of the airflow during supersonic flight to a subsonic Mach number of about 0.8. Although designed to fly at Mach 3, we never flew the airplane beyond Mach 2.55. The aircraft had delta wings that were hinged so that the tips could fold down to take advantage of a phenomenon called compression lift. The long forward fuselage sported two horizontal control surfaces called canards. The XB-70 was painted gloss white overall, and looked like a great swan in flight.

I was paired up with two instructor pilots, Fitz Fulton (now with NASA) and Air Force Col. Joe Cotton. Both had a great deal of experience in the XB-70. Like Fitz Fulton, Cotton was a joy to work with. It was obvious to me why these gentlemen had achieved positions of high regard and responsibility in the flight test community. Joe was aware that his Air Force career was in jeopardy because of his role in the accident with Ship Two. His chances of attaining general officer rank and continuing his military career were slim at best. There was never a hint of bitterness from Joe during the NASA-USAF program, and my high opinion of him only grew. He eventually retired from the Air Force and went on to a second career as a test pilot for United Airlines. Fitz and Joe shared their expertise with me, and my checkout period went smoothly. They knew the airplane as well as anyone could, and quickly got me up to speed.

I flew my first four flights in the right seat as co-pilot with Fitz, beginning on 22 June 1967. Although it was checkout time for me, we also gathered important research data on each flight. The right-hand (co-pilot's) seat was an excellent place to learn about the aircraft, especially systems operation. The XB-70, as large as it was, had no flight engineer to manage the systems. The co-pilot served that function and several others. Of course, there were flight

Mallick took this photo of the XB-70 during a chase mission. Safety chase pilots fly in tight formation to observe the test aircraft closely.

controls on both sides and I accrued as much flying time as possible during each mission. The fuel and engine inlet systems were manually controlled. I had a heavy workload, manipulating all the switches, levers, and controls to keep this marvelous piece of machinery in the air. After completing the flight test card, I could get a little more exposure to the controls and flying characteristics, especially in the landing pattern.

During my fifth flight in the XB-70, on 13 February 1968, I had the honor of occupying the left (pilot's) seat with Joe in the right. It was my formal check flight as first pilot in the XB-70. I had flown many missions with Joe in the TB-58A and I was comfortable with him. We knew each other's flying ability and, if he thought half as much of me as I did of him, we were in good shape.

With the cockpit so high above the ground, taxiing the XB-70 was an unusual experience. The cockpit was located well forward of the nosewheel, and the main wheels were even farther to the rear. The forward section of the fuselage was like a long cantilevered beam with the crew compartment located out near the end. The cockpit windscreen had two positions, one for the landing and takeoff, and one for high-speed flight. This provided adequate pilot visibility in both modes as well as lower drag at high speeds. A big problem while taxiing was guessing the location of the landing gear relative to the taxiway. This made for interesting sensations during a turn. We had to wait until the cockpit was over the edge of the taxiway toward the desert before we actually initiated the turn. This insured that the main gear did not drop off the taxiway

USAF photo

Joe Cottton (left) and Mallick pose in front of a navigation chart of western U.S. test ranges. Cotton served as the primary Air Force test pilot during the joint USAF-NASA XB-70 research program.

on the inside of the turn. I observed this during my time in the co-pilot's seat and learned to taxi from that side. It was different from any other aircraft, and difficult to get used to.

As we left our parking spot, aligned perpendicular to the taxiway, we only taxied a few yards forward before starting our turn. Unfortunately, the nosewheel didn't provide adequate steering torque at heavy weights, and low taxi speeds. There just wasn't enough steering torque available to turn the aircraft until we had built up some speed. This made it exciting when coming out of the parking area because we needed a significant amount of engine thrust to overcome the inertia of the parked aircraft. We had to leave the power on until achieving a relatively moderate taxi speed. Of course, as we applied rudder input for the turn, we prayed that the system worked and the big aircraft swung around. If it hadn't, there would have been no stopping the aircraft before it crossed the taxiway into the dirt beyond. Fortunately, the technique worked consistently, and I always heaved a sigh of relief as the nosewheel steering brought the aircraft in line on the taxiway.

Seated so high above the ground, it was difficult to judge taxi speed. Because we had no groundspeed readout in the cockpit, we relied on escort vehicles to call out our speed occasionally on the radio. We instructed them to give us a call when we exceeded 25 miles per hour. It was hard to stop the aircraft once we had some speed, so we didn't want to go too fast. Also, when coming to a complete stop, the pilot had to be smooth in applying the brakes in order not to lock the wheels.

Lining up on the runway was similar to the turn onto the taxiway, but much easier. At Edwards, a wide taxiway leads to Runway 04-22, the hard-surface runway on the main base. The runway itself is 300 feet wide and nearly 15,000 feet long. We use the same technique on lineup as we had turning onto the taxiway. Once the cockpit position passed way beyond the runway centerline, we applied a large rudder input to activate nosewheel steering, and swung the aircraft back toward the runway centerline. The runway was wide enough that it was not essential to have the nosewheel exactly on the runway centerline, but we aligned it as close as possible.

Once we had the XB-70 positioned on the runway and held with normal brakes, we completed the pre-takeoff checklist. We advanced the throttles and selected minimum afterburner on all six engines. We then released brakes and pushed the throttles full forward to select maximum afterburner. At this point, the engines produced a total thrust of 180,000 pounds. We released brakes and the aircraft accelerated on its takeoff roll. Even with so much thrust, the initial acceleration was slow due to the overall weight of the aircraft. As our speed increased from around 80 to 100 knots, we began to feel runway roughness. The XB-70 was a very flexible aircraft and the cockpit responded to any surface irregularities, bouncing up and down on the end of the fuselage. Runway 04 always gave us the roughest ride at about 3,500 feet from the start of our takeoff roll.

As the aircraft passed through 175 knots indicated, the pilot pulled back on the yoke. The aircraft responded, with the nose rising slowly. As the nose approached a position where the cockpit windscreen lined up with the outside horizon, the pilot eased off on the stick and held the aircraft's pitch angle steady. As the XB-70 accelerated to a speed of 215 knots, the cockpit was about 40 feet above the ground while the main landing gear was still in contact with the runway. At this point, the main gear lifted off the runway and the XB-70 was flying. It was quite an experience, and it always seemed like a long time between achieving takeoff attitude and reaching flying speed. It was important to avoid over-rotation of the nose prior to takeoff, which reduced acceleration and greatly increased the takeoff roll.

On one occasion during the Air Force XB-70 test program, I was waiting for takeoff in a NASA F-104 behind the XB-70 and observed an over-rotation on takeoff. It was a warm day, and the pilot must have pulled back too far on the yoke. The combination resulted in an excessive takeoff roll that caused the

NASA photo E67-16695

The XB-70 resembled a great swan on takeoff. Although cumbersome to maneuver on the ground, it was very graceful in flight.

XB-70 to disappear in a cloud of dust on the departure end of the runway. I thought initially that it had crashed, but the great white bird emerged from the dust cloud intact. It seemed as if it was clawing to stay in the air, and it succeeded. Van Shepard, the North American pilot, was at the controls with Fitz as co-pilot. Fitz remarked to me later, "I was never so glad to get the gear up on that aircraft in my life." Van was a good pilot. He had just over-rotated the big bird and it wouldn't stand for that.

Once the XB-70 was airborne, it did not seem quite as cumbersome as it did on the ground. It flew well and responded nicely to the pilot's control inputs. It handled very well throughout its flight regime, with the exception of dampers-off flight around Mach 2.3 to 2.6. In this area, the XB-70 had a lateral-directional oscillation that was difficult for the pilot to dampen manually. The frequency was such that the pilot could only attempt to dampen out the unusual "Dutch Roll" at one point during each cycle. Any attempt to fight the aircraft too quickly resulted in a pilot-induced oscillation (PIO) that only aggravated the situation. Of course, normal flight operations were not conducted with the dampers off, but there was always the possibility that the dampers could

NASA photo EC68-02052

Fitz Fulton (right) confers with Mallick about an XB-70 research mission. The two men alternated pilot and co-pilot duties.

malfunction. Therefore, we conducted dampers-off flights to develop techniques for dealing with emergencies.

Flying at higher Mach numbers gave us a chance to study precise pitch control in order to hold altitude and airspeed. At higher speeds, a very small change in nose pitch position resulted in sudden altitude and airspeed changes. This was characteristic of any high-speed aircraft, so our pitch control research had numerous applications to aircraft development and pilot training.

In the landing pattern, the XB-70 handled like the large aircraft it was. Roll response was good, similar to a smaller aircraft. Fortunately, roll inertia was not large relative to the roll control. In pitch, it acted more like a large aircraft, such as the B-52. But even with these disparities in the two axes, the XB-70 flew well in the pattern and was easy to land. The pilot's greatest challenge during landing was guessing where the main wheels were relative to the runway. After nine flights in the XB-70, I still couldn't estimate the height of the main gear above the runway any closer than 10 feet. At times, the chase aircraft had to call out the last few feet prior to touchdown. The XB-70 pilot established a reasonable rate of descent for touchdown with the main landing gear about 50 feet above the runway, then held that sink rate until the wheels touched the concrete. Fortunately, the XB-70, like most delta-wing aircraft, had a strong ground effect that assisted the pilot in making a smooth touchdown. Ground effect is the change in airflow around the aircraft when it approaches close proximity to the ground, about one wingspan above the runway. It acts like a cushion to reduce the rate of descent.

After landing, the pilot faced the problem taxiing back to the parking area. Now, however, there was much less fuel onboard. With the lighter aircraft weight, nose wheel steering was much more responsive, and parking the aircraft was much easier than taxiing for takeoff

On Fire!!!

Whereas I usually flew the XB-70 with Fitz, Ted Sturmthal usually flew with Joe Cotton. The two crews alternated flights. Fitz made some flights with Ted and I made one with Joe. In that way, we always had a senior XB-70 pilot onboard. There was never a question that Ted and I could have operated the aircraft together. It just seemed smarter to have a mixed crew, and I had no heartburn about that. It was "good headwork," as we liked to say. I flew a total of nine flights in the XB-70 over a period of a year and a half. Ted flew nine research flights and helped Fitz ferry the aircraft to the U.S. Air Force Museum in Ohio when the program was finished. On missions where I didn't fly the XB-70, I flew as a chase pilot in the Air Force TB-58A.

On 2 November 1967, I was flying in the co-pilot's seat of the TB-58A and Fitz was in the front. Joe and Ted were flying the XB-70 to gather inlet data and study longitudinal handling qualities. We couldn't follow the XB-70 during the entire mission due to performance differences. The TB-58A could not keep up with the XB-70 after it passed Mach 2.0, and the test card called for the XB-70 to exceed Mach 2.5. On this particular mission as in others, the TB-58A was scheduled to fly an inner pattern as the XB-70 went on its much longer, higher, and faster profile. That way, if the XB-70 had a problem anywhere along its route, the chase aircraft would cut across the flight track to intercept it and escort it to a landing. On this flight, Fitz and I dropped in behind the XB-70 near Las Vegas, Nevada as it headed north. Fitz noted that we were running a little low on fuel. On each flight, we monitored a time-position-fuel plot to keep track of the overall progress of the flight and fuel remaining. Fitz slowed the TB-58A, and we went into a fuel-conservation mode in order to remain in the air to cover the XB-70's landing.

All went well with the XB-70. When he came back into the Edwards area from the north, we joined up with him and visually inspected the XB-70. This was standard operating procedure for the chase crew. Fitz then asked Cotton about his fuel levels and Joe replied, "We are in fine shape, Fitz". "Joe," said Fitz, "we are a little low on fuel. Would you drop your landing gear and let us check them? And then, we would like to land first." In the past, the XB-70 had

Flying the Heavies

Courtesy of the Air Force Flight Test Center History Office

The TB-58A often served as a chase plane for the XB-70. Of course, during high-speed, high-altitude missions, the TB-58A could not keep up with the XB-70 throughout its mission profile.

trouble with the extension of its landing gear, so the chase crews always made a visual check before landing. It was not standard for the chase to land before the XB-70. Although Fitz had not been calling off fuel readings to me, I knew that we were low. Joe dropped the gear and we looked it over. There were no problems. After a final visual check, we departed the XB-70 and called the tower for a straight-in landing to Edward's Runway 04.

Fitz made a nice approach and a smooth landing, but as we rolled past mid-field we both heard a loud pop. "That sounded like a tire blowing out," Fitz commented, "but it doesn't feel rough." I acknowledged and I too was at a loss as to what the problem was. As we approached the eastern end of Runway 04, we heard the tower tell "001" (the XB-70) to go around. Fitz, in turn, who had the TB-58A well under control and near turnoff speed, called Joe and said, "001, we are clearing the runway shortly and you should be able to land behind us." Edwards Tower immediately responded, "No, 001, go around. The B-58 is on fire."

This was the first we knew that we were on fire. The tower hadn't bothered to tell us. Bright orange flames were shooting out the back of the No. 3 engine, where we couldn't see it during the rollout. Apparently, the pop we heard was the fire starting on the right side. Fitz alerted me, and the Air Force crew chief in the third seat, to prepare to abandon the aircraft once we stopped clear of the runway. As we turned off onto the taxiway, the fire was obvious. Flames were leaping high on the starboard side of the aircraft, above the cockpit level. There wasn't any question about which side of the aircraft we would use to evacuate. I unstrapped my parachute harness, oxygen and communications cords, and attachments even before Fitz stopped the aircraft. He set the brakes, shut all four engines down and we exited the aircraft immediately. Later study of NASA long-range optical coverage indicated that all three crewmembers were on the ground and running away from the aircraft within 20 seconds of stopping. It was a quick egress!

We watched with amazement as the Air Force fire crews fought the blaze for some time before extinguishing it. The TB-58A was saved. After the fire, the maintenance crew discovered that a four-inch clamp on a fuel filter in the Number 3 engine had worked loose and leaked. That leak resulted in our low fuel readings. We had no way of seeing the fuel trailing behind our aircraft, and apparently the tower crew failed to observe it during our landing approach. After we touched down, the fuel-air ratio was just right for ignition. Thus, the bang after landing and the ensuing fire. It could have turned out much worse. After the fact, I was impressed when I recalled that I had felt no fear or panic. I remember looking out the right-side cockpit window and seeing the flames. "That's not the way out," I thought. As soon as the canopy opened, I jumped over the left side and dropped to the taxiway. I ran away from the aircraft as fast as I could. Some distance away, I stopped and looked at my fellow crewmembers, then back to the burning airplane. It had been a close call!

End of the Program

The NASA-USAF XB-70 flight test program began in November 1966. A total of 22 research flights were performed between April 1967 and December 1968. The XB-70's flying career ended with a subsonic delivery flight to the Air Force Museum at Wright Patterson AFB in February 1969. I flew nine missions in the aircraft, five as co-pilot and four as pilot. I went through ground schools and flight checkouts in two other aircraft, the B-58 and B-52, in preparation for flying the XB-70. I would have done all of it for just one flight, so I was happy to get nine. I wished I could have had at least 20 flights. The XB-70 was expensive to maintain and operate. After the 22 flights, NASA had determined that they had obtained as much flight data as they could afford from the program, and brought it to an end. I was happy to have been part of it and truly enjoyed flying this large, impressive airplane.

Other Test Programs (1965-1970)

There were a number of other programs at the Center during this period. They included continuing X-15 research flights, lifting body tests, and a number of other smaller aeronautical flight research programs. NASA flight research operations were unique and demanding, and they were not achieved without some cost.

In 1967, Bruce Peterson was injured in a crash involving the M2-F2 lifting body during a landing on the north end of Rogers Dry Lake. Lifting body aircraft, in general, had some undesirable lateral-directional flight characteristics when flying at low angles of attack (AOA) at high speeds. The lightweight M2-F1, that I had the opportunity to fly, was no different. "Don," Milt Thompson warned me during my checkout, "don't even screw around with the rudder pedals." At low AOA, a small sideslip angle created a large rolling moment, even larger than normal aileron input could handle. In addition, if the pilot attempted to dampen the oscillation, the disparity between the aircraft's and pilot's responses resulted in PIO. The best technique to mitigate an unwanted roll-yaw oscillation involved increasing the vehicle's angle-of-attack by pulling back on the stick to allow the vehicle to enter a more stable region of flight.

On 10 May 1967, Bruce was a little high on energy during the final approach. He was also flying in the critical low AOA range and soon experienced a lateral-directional divergence. The stubby, wingless vehicle began to swing from side to side before Bruce could increase his AOA. Applying the proper inputs, he quickly brought the beast under control, but the maneuver had carried him away from his planned approach path to lakebed Runway 18. Too low to correct, Bruce chose a landing spot devoid of any visual reference markings. He had no choice since his gliding vehicle had no power for a go-around. The lack of visual cues made it impossible for Bruce to accurately estimate his height above the lakebed. He was further distracted by the appearance of a rescue helicopter, hovering over the landing area. "Get that chopper out of the way," he called. Now very low to the ground, Bruce flared and lowered his landing gear. It was too late. The gear was not fully down and locked. The M2-F2 struck the ground, bounced, and rolled across the lakebed. As the dust cleared, horrified observes saw the vehicle on its back. The canopy and parts of the tail fins and landing gear were strewn across the hard-packed clay.

On the days that I was not directly involved in research or chase flying, I normally observed other test operations. On this day, I watched the M2-F2 landing from the roof of our main office building. I could not believe that anyone could survive an accident like this, and immediately headed down to the Flight Operations office. I watched as Air Force and NASA crash rescue personnel responded to the accident. Stunned rescuers found Bruce still alive, but badly injured. He even tried to sit up in the gurney as he was carried toward an ambulance.

My boss sent me to the Edwards hospital to check on Bruce following his emergency surgery. It was not a pleasant assignment. I expected to arrive at the emergency room only to find that Bruce was dead. When I was admitted to the treatment area, I couldn't believe my eyes. Bruce was sitting up! His head was almost completely wrapped in bandages and swollen to almost twice its normal size. He was conscious and trying to talk to those around him. I couldn't believe it! They almost immediately loaded Bruce back into an ambulance, drove him to the flight line and directly onto a transport aircraft. They then flew him to March AFB, which had a much bigger hospital with more facilities and care available. Several days later, he was moved to the hospital at the University of California, Los Angeles (UCLA).

Bruce survived. He lost an eye after the accident, but this was due to a staph infection that he contracted at the UCLA hospital during his recuperation. He remained at FRC as the director of safety and continued to fly with NASA and the Marine Corps Reserve. Bruce was fortunate in that he survived an accident that would normally have been fatal.

On 25 July 1967, Hugh Jackson, experienced an inflight explosion while flying the twin-engine F-4A jet fighter. Hugh, who had been at FRC for just over a year, made an emergency landing on the north lakebed with no further problems. Inspectors found a tremendous hole in the right wing that Hugh could stand up in without touching any of the surrounding surfaces. Apparently, a spark from an exposed electrical wire ignited fumes in an empty wing fuel tank. There was no evidence of soot or explosive residue, but milled wing skin (nearly an inch thick in some places) was blown away. Hugh was lucky. No hydraulic lines, landing gear, or flight controls were affected. Fortunately, the main wing structure remained intact, so Hugh could bring the airplane home safely.

Maj. Mike Adams was not so lucky. An Air Force pilot assigned to the X-15 program, Mike was killed in a crash of the number three aircraft on 15 November 1967. It was Mike's seventh flight, the one on which he would earn his astronaut wings by exceeding 50 miles altitude. He reached a maximum speed of Mach 5.2 and an altitude of 266,000 feet. Arcing gracefully over to reenter the atmosphere, Mike was distracted by an electrical transient that degraded his control system. Plagued by control problems, and perhaps vertigo, he failed to notice a gradual drift in the vehicle's heading. The X-15 turned sideways, then backward, and continued to turn while still at hypersonic speed. Adams radioed that the vehicle seemed "squirrely." Finally, realizing what had happened, he called the control room again. "I'm in a spin." There was no panic in his voice, but he must have known it was a bad situation. There was disbelief in the control room. A hypersonic spin was new territory. "I'm in a spin," Mike said again from the edge of space. Mike fought the spin as the X-15 plunged toward Earth at Mach 4.7. Somehow, he managed to recover from the spin, but entered an inverted dive. He was still high enough to regain control, but the adaptive flight control system began a limit-cycle oscillation. The X-15 began to pitch up and down wildly. As the vehicle plunged into the denser atmosphere, structural loads increased dramatically. Dropping through 62,000 feet and a speed of Mach 3.93, the vehicle broke apart. The forward section slammed into the desert near the old mining town of Johannesburg. Mike was dead.

It was the only fatality of the 199 flights of the X-15 program. Considering the high-risk nature of the project, that was quite amazing. I never flew with Mike, but I knew and liked him. On several occasions, I shared a drink with him at the Friday night get-togethers that were common in the days of the X-15. Mike was sort of laid back and unassuming. He sometimes seemed surprised that he was flying something like the X-15. His loss was the kind that, as a fellow pilot, I tried to accept and then press on. As I have said before, it always took a little of me with them. They were a part of my life that is now no more.

On 6 May 1968, Neil Armstrong ejected safely from a LLRV at Ellington AFB near Houston, Texas, while training for his lunar landing mission. Neil was one of the first astronauts to check out in the LLRV, and when he had to eject, everyone was concerned about the safety of the vehicles. I was sent to Houston to serve on the accident investigation board because of my earlier flight test experience on the LLRV.

One thing that I, and others at FRC who had tested the LLRV, found obvious was that this was still a research vehicle. Even though NASA now called it a trainer, the LLRV was not a production-type vehicle. It was a sophisticated machine that had been tested for a few years under controlled conditions, and there were many hazards to be aware of. It was not like the T-38 jet aircraft that the astronauts used for proficiency training. We at FRC had tried to pass on all the information that we had on the LLRV, which included many warnings about its limited flight envelope. We had checked out two pilots at Edwards as instructors for the operations at Houston. A design flaw that we had lived with at Edwards for several years, and were unaware of, came out to bite Neil.

Neil had been flying in the Lunar Simulation Mode when he tried to switch the vehicle back into the VTOL mode, a normal procedure that we had done dozens of times. Neil, however, had not moved the "lift stick" (peroxide lift rocket control) to the full-down position. There was still sufficient peroxide remaining for the attitude control rockets, but when the lift rocket peroxide level reached zero, the helium source-pressure gas vented out through the inactive lift rockets. This was the same source-pressure required for the attitude control rockets that were needed for both modes of the vehicle, lunar simulation or VTOL. As the helium source was depleted, Neil lost attitude control and was forced to eject at low altitude. His parachute carried him to a safe landing near the burning wreckage of his vehicle.

What happened to Neil had happened numerous times when new aircraft were put into production or squadron service. Even after several years of testing, undiscovered hazards sometimes emerged. This flaw had slipped by in the design phase and had never been discovered during testing, but Neil found it quickly in operational use. We, and especially Neil, were fortunate that the Weber ejection system functioned properly. As it turned out, two other LLTVs were also lost during astronaut training, although not for the same reason. In my opinion, the LLRV/LLTV was still a research vehicle, not a trainer. It was a complicated, demanding machine to operate, but provided excellent experience for the astronauts in their preparation for the moon landing.

The late 1960s took their toll of aircraft and pilots. It seemed as if the bad periods came in cycles, for some reason. Sometimes, a period of one or two years resulted in an unusual number of aircraft accidents. It was a part of test flying that we didn't like, but accepted grudgingly. Risk is inherent in flight test.

Chasing the Northern Lights in a Convair 990

In late 1968, I was assigned to a transport aircraft study with Fitz Fulton. Fitz and I seemed to be the "big airplane" test pilots for NASA FRC, following the XB-70 assignment. I didn't mind a bit, because I realized at this stage of my career that any aircraft could be an exciting challenge for a pilot. I was blessed by the opportunity to experience flight in helicopters, fighters, bombers, transports, light civilian aircraft, and sailplanes. Fitz was about six years my senior, and he was an excellent test pilot with a wealth of experience. What better person could I have asked for to share a flight program with?

For this project, Fitz and I "borrowed" a Convair CV-990 (nicknamed "Galileo") from NASA Ames Research Center at Moffett Field, near San Francisco. The program was designed to study the low-speed handling characteristics of the CV-990 during the final approach and landing phase of flight. We flew a comprehensive stall study program with the aircraft in all configurations. Additionally, we made landing approaches and touchdowns as much as 15 knots below the speeds recommended in the flight manual, in order to determine reasonable limits for pilots to make safe landings. All these tests were a little hard for me because it seemed that we were abusing the aircraft to gather data. We were, in fact, but for good reason! We pushed the aircraft to limits where we predicted that even a skilled pilot would break the airplane in one of three landings. That seemed to satisfy the engineers, and they did not ask for any more mistreatment of the aircraft. I recall some of the "clean configuration" stalls that caused the aircraft to shudder, buck, and shake to the point of loosening rivets. Fitz and I talked the engineers into canceling a number of the repeat points (that engineers seem to like) just to protect the old CV-990 from being shaken to pieces.

After a six-month program, the CV-990 was returned to Ames at Moffett Field and reassigned to its primary mission of airborne science studies. The aircraft was equipped with about 25 scientific stations and experiments to study the atmosphere, weather, the Earth, and the heavens. A number of universities

Flying the Heavies

NASA photo E70-21667

Fitz Fulton and Mallick "borrowed" a Convair CV-990 from the NASA Ames Research Center in order to study low-speed handling characteristics during final approach to landing.

conducted studies and provided scientists for these programs. When the CV-990 was deployed on a scientific mission, it could be "on the road" for months and fly all around the world. This created a pilot drain in the Ames Flight Operations office that was normally busy conducting VTOL and V/STOL flight research. For this reason, and the fact that Fitz and I were now qualified in the CV-990, we were invited to participate on some of the scientific research missions.

When I was asked to fly the CV-990 with Ames pilot Fred Drinkwater in December of 1969, I jumped at the chance. I had no idea what beauty and astounding sights were in store for me as one of the pilots on a scientific study of the Aurora Borealis, or "Northern Lights" as I knew them as a kid in Pittsburgh. We were based at Fort Churchill, Quebec, on the western edge of Hudson Bay. Flying in and out of Fort Churchill in December was an experience. All of the crew and scientists were outfitted with cold-weather gear - parkas, insulated pants, and gloves. The temperatures dropped to 35 degrees below zero at night, and were not much warmer during the day. The CV-990 was sheltered in a relatively small hangar with a special ramp inside to tow the nose wheel up in order to lower the tail for clearance under the hangar door opening. This was necessary in order to perform maintenance on the aircraft in a heated area.

Prior to flight, which was always at night, the hangar doors were opened on both sides and the aircraft allowed to "cold soak" (chill down) to outside temperature prior to rolling it out on the ramp. This prevented snow from freezing on the surface of the aircraft. When the CV-990 was pre-chilled, the snow just bounced off. At such extremely low temperatures, very little snow formed in the air, but the ground was covered with snow left over from winter storms. The wind blew constantly, causing visibility problems on take off and landing. This was the first time that I had ever flown in extremely cold conditions, and I was impressed with the planning and special precautions necessary to permit operations in this environment. Once the engines were started, the aircraft's normal heating system was adequate to warm up the cabin and the crew removed their heavy winter gear. After a five-hour mission, we returned our warm airplane to the same "ice box" environment. After parking and shutdown, we bundled ourselves back into the bulky clothing that was required for even a few minutes outside.

A flier has the opportunity to witness the world from a position unlike any other at altitudes of 8 or 9 miles above the ground. I recall the beauty of maneuvering between high (50,000 feet plus) thunderstorms in southern Florida in the summer. The power to observe and safely admire the majestic storms always impressed me. If a flier entered one of these majestic storms, on purpose or by accident, the results could be terrifying, if not catastrophic. Flying at high altitude, at night, in perfectly clear weather in the area of the North Pole was equally majestic. As the CV-990 cruised on autopilot at 41,000 feet, the air was smooth as glass and the cockpit lights were dimmed to a very low green. The normal hum of the aircraft instruments, fans, and other motors provided a comforting background, and the inertial navigation systems told us within a tenth of a mile of our location even though we were but a speck in the vast night sky. The stars outside the cockpit windows were magnificent, so bright and clear. Their blue-white radiance stretched from horizon to horizon. There were no lights on the ground. I felt as if I could see over the edge of the Earth. When flying over land areas that are heavily populated, ground lights merge confusingly with the lower stars. Aircraft have actually crashed when the pilot mistakenly identified a ground light as a star and became disoriented. Not so in northern Canada, near the North Pole. The few ground lights we saw during the long, dark hours belonged to the Distant Early Warning (DEW) Line radar stations. They were easily identified as they were in a long line and spaced widely apart.

An aurora is a luminous atmospheric phenomenon that occurs close to the poles. Near the North Pole, it is called Aurora Borealis and in the south it is known as Aurora Australis. It occurs when high-energy particles from the Sun are expelled into the Earth's magnetosphere by the Van Allen radiation belt.

This results in a fantastic light show as rapidly shifting curtains and columns of light dance across the sky in myriad colors.

The scientists selected Fort Churchill as our mission base because it is an active area for the aurora late in the year. The phenomenon has a "lifespan" that varies, depending on the number of protons and electrons interacting with gas molecules in the atmosphere. The electromagnetic discharges form, grow, reach a maximum strength, and then slowly lose their energy. On this night, we had flown much closer to the North Pole than usual because we had not had any luck around the Fort Churchill area on prior flights. At first, there wasn't a great deal of auroral activity around the Pole, and we set course back toward Fort Churchill. After about 30 minutes on a southerly course, we noticed a small but bright spot on the horizon directly ahead. As we flew on, the spot grew larger and brighter before taking the shape of a rectangular screen, similar to the "drive-in theater" screens that were popular in the days of outdoor movies. As we flew on, the rectangular screen grew higher and wider, but was always rectangular in shape. The scientists' interest grew as they came to the cockpit in twos and threes to view the phenomenon. It glowed in yellow, red, and green, and there were shimmering vertical lines and waves.

It was fantastic! As we flew closer to it, the "screen" seemed to stretch from horizon to horizon. We watched in awe as it burst into dozens of brightly-lit streamers. As we strained to look behind us through the cockpit windows, we found ourselves in the center of a "bowl" of auroral activity. The show continued for about five or ten minutes. The view was so breathtaking that we hardly spoke to one another. Slowly, after about 45 minutes, the heavenly colors began to fade. Either because we had passed the active area, or because it had lost its energy, the aurora's intensity began to subside. Our scientific instrumentation was humming away as it took the data. The scientists occasionally took quick breaks from their experiment consoles to dash forward and look out at the view. They also took photographs through the cockpit windows and the results were impressive, but they couldn't match the pictures that we recorded in our minds that night. I'm sure that I will never see a sight like that again, and I thank God for the chance to see this one. The cockpit of the CV-990 provided me with a ringside seat to witness this wonderful phenomenon first-hand. Few people have ever had that opportunity. I was glad that I had volunteered for the mission, and it made all of the cold-weather operation worthwhile.

After several weeks at Fort Churchill, we departed for four-day deployment to Bodo, Norway. Planning for the flight was impeded because we found it difficult to obtain up-to-date weather information for the Bodo area. Limited information was available at Fort Churchill and it was always out of date. After gathering every bit of weather data available, Fred Drinkwater and I filed our flight plan for Bodo and used another airport in southern Norway as an alternate. On the way to our destination, we took some time for research. The scientists had all of their instrumentation running, but we did not encounter

NASA photo E72-25265

The Convair CV-990, nicknamed "Galileo," served as an airborne science laboratory. In 1969, the aircraft carried scientists to the northern latitudes to observe the Aurora Borealis.

Flying the Heavies

any outstanding auroral activity during the flight. About two hours west of Bodo, we obtained a current weather report through the air traffic control network, and were informed that Bodo airport was closed due to snow and ice storms. Because of the fast-moving storm front, Fred and I pressed on toward Bodo, knowing that we could fly south to the alternate if required. About an hour west of Bodo, we were informed that the storm had passed over and the airport was open. Icy runway conditions promised to make braking difficult. As airplane commander, Fred made the decision to land at Bodo despite the conditions. Fred was a competent pilot, and I had complete confidence in his judgment. We alternated flying the left seat, but he made all of the decisions affecting the mission.

As we broke out of a 2,000-foot overcast, we spotted the lights of Bodo. The storm had moved north and the ground shimmered white with fresh snow. Fred flew an excellent approach, precisely on speed and glide path, and we touched down as planned. The runway was rough with ice, snow, and slush. Fred deployed the spoiler and carefully applied reverse thrust. The CV-990 stopped gently at the end of the runway. The aircraft lumbered along the unplowed taxiway to the parking area. Because we were flying around the top of the world where the time zones are compressed, we crossed five time zones during the five-hour journey! Jet lag was a serious problem. I was never able to convert my body's internal clock to local time, and I never slept more than two to three hours at a time while in Bodo.

The next night, it was my turn to fly left seat for a research mission. As we approached the aircraft on the now frozen ramp, I wondered if I would even be able to taxi the aircraft, let alone get it into the air and fly. Some of the scientists slipped and fell on the ice as they approached the aircraft. It wasn't reassuring. Fortunately, the runway had received more attention. Although covered with snow and ice, it was not quite as bad as the ramp and taxiway. With judicial use of throttle, nose steering, and a little reverse thrust at times, taxi, takeoff and flight were uneventful.

The people in Norway were wonderful. The buildings and shops were clean and bright, and the weather in Bodo was relatively warm. The temperature averaged around 20 degrees Fahrenheit, which was much warmer than the minus-35-degree temperatures in Fort Churchill. I remember seeing Norwegian ladies out shopping with children, all bundled up in their baby buggies. The ladies left their children outside of the stores as they shopped inside. This was to keep the children, completely bundled up and warm for the outside, from becoming overheated inside the stores. I always marveled at how safe the town seemed and how little crime they had. It was impressive.

We returned to Fort Churchill after four days. Although the weather was much colder, the runway was cleaner and the most we had to contend with on landing or takeoff were the 30-knot crosswinds that Fred and I had learned to handle. We returned to Moffett Field a week later and my winter flying experience was over. I had enjoyed it very much. It was a big change from my normal routine.

9

Triple-Sonic Blackbird

High Speed Flight Research

In the late 1960s, the NASA Flight Research Center seemed to be moving away from high-speed flight research. Most of the ongoing high-speed programs just happened to be phasing out at the time. The X-15 made its final flight on 24 October 1968, and the last research flight of the XB-70 was on 17 December. At this time, there were a number of lifting body aircraft under evaluation. The M2-F3, HL-10 and the X-24A were still flying, but they were primarily designed to evaluate the re-entry and landing characteristics of a space shuttle type vehicle. These rocket-powered vehicles were capable of supersonic flight, but there were fewer programs involving supersonic jet aircraft.

In the early 1960s, Ames Research Center supported the development of a family of secret spy planes designed by Lockheed Aircraft Company. The twin-jet, supersonic aircraft included the A-12, YF-12A, and SR-71 (collectively known as the Blackbirds for their overall flat black exterior finish). NASA technicians tested models of the various configurations in high-speed wind tunnels at Ames. They determined that the airframe would have a design speed of Mach 3.2 in cruise, making it the fastest air-breathing airplane in the world. Initially, the existence of these aircraft was highly classified. On 29 February 1964, the YF-12A became the first of the Blackbirds to debut publicly. Within a year, NASA officials expressed an interest in using the airplane for high-speed flight research. Permission was not forthcoming, for security reasons, but NASA persisted.

In 1967, NASA officials approached the Air Force to request some of the SR-71 test data to use for comparisons with the XB-70 flight test data. Eventually, the Air Force agreed to let NASA technicians install instrumentation on one of their test aircraft. Although NASA officials wanted to obtain one or two SR-71 aircraft to be flown by NASA crews, they had to be satisfied for the moment just working with the Air Force.

An event on the other side of the globe ultimately assisted NASA in obtaining several of the high-speed aircraft. The Russians had introduced a Mach 3 fighter, the MiG-25, into their inventory as a response to the perceived threat of the B-70 (which never materialized due to cancellation of the bomber). The Soviets succeeded in building an airplane that could reach Mach 3, but only in dash, not cruise. The Air Force countered with a request for a Mach 3 interceptor, based on Lockheed's then secret A-12 spy plane. Three prototypes, designated YF-12A, were built. Capable of firing missiles at airborne targets,

Triple-Sonic Blackbird

NASA photo E71-23131

The Lockheed YF-12A was originally designed as an interceptor. NASA used it as a research a platform to study propulsion, heating, loads, aeroelasticity, and flight dynamics of a supersonic cruise aircraft.

the aircraft underwent extensive tests between 1963 and 1965. The aircraft also set several speed and altitude records, shortly before being placed into storage when the Department of Defense decided not to put the aircraft into production. The first YF-12A was badly damaged in a landing accident at Edwards in 1966. The two remaining aircraft were soon placed in storage.

In 1969, the Air Force agreed to let NASA fly the "mothballed" airplanes in a joint NASA/USAF research program. The Air Force provided the airplanes (serial numbers 60-6935 and 60-6936), personnel, ground support equipment, and facilities. NASA agreed to pay the operational expenses, using funding left over after termination of the X-15 and XB-70 programs. The Air Force portion of the project consisted of research to explore the tactical performance and support requirements of an advanced interceptor. This included radar intercepts of flying target aircraft (such as the F-106 and F-4), demonstration of a semi-autonomous concept for a Mach 3 interceptor, and assessment of the feasibility of a visual identification maneuver against a supersonic transport-type aircraft.

The NASA phase of the program included research into propulsion, aeroelasticity, heating and loads, and flight dynamics of a supersonic cruise aircraft. Ship 936 handled the bulk of the Air Force flight tests while 935 mainly served the NASA research effort. We also used 936 for some NASA checkout flights, and the Air Force used 935 for some of their tests early in the program.

The Air Force supplied some very talented pilots, mechanics and technicians with Blackbird experience to man the program. When we started, in 1969, Maj. William J. "Bill" Campbell was the operations officer for the SR-71/F-12 Test Force at Edwards, under Col. Joe Rogers. Bill, Joe, Col. Hugh "Slip" Slater, and Lt. Col. Ronald "Jack" Layton served as the Air Force project pilots. In 1970, Bill was promoted and took over as Test Force director. He was a fine pilot and manager, and a real gentleman. Fitz Fulton and I were designated the NASA project pilots. Since we had recently finished the XB-70 program, we had experience with high-speed supersonic aircraft. It was such a unique opportunity that I did not feel so bad about missing out on the X-15 program. The assignment turned out to be one of the longest and most productive flight research programs of my career.

Flight test engineers Victor W. Horton and William R. "Ray" Young served as our "backseaters." They kept track of our test points during the flight and recorded data points. They also managed the aircraft's fuel consumption and served as navigators. Although Fitz and I flew with both Ray and Vic at one time or another, we gravitated into two teams for the Blackbirds: Fitz/Vic and Don/Ray. We flew the majority of our flights with these two crews. Vic joined NASA in 1959 and worked on the paraglider and lifting body programs. He had some flight training years earlier, but for some reason never completed it. He did, however, love airplanes very much. Vic studied engineering and ended

NASA photo EC72-02982

NASA YF-12 crewmembers included (from left to right) Ray Young, Fitz Fulton, Don Mallick, and Vic Horton.

up as close to flying them as he could. He passed away a few years after retirement. Vic had lived with a kidney problem for years and with the knowledge that it was not curable. He showed the same courage facing this problem as he did in flying.

Ray, a former Air Force jet pilot and engineer, came to NASA FRC in 1962. He initially worked on the LLRV before moving on to other projects, including the X-15, XB-70, and F-111. I attribute a large percentage of my Blackbird accomplishments to Ray. He was a very important part of the team and contributed greatly to our success.

I held a great respect for both Vic and Ray, but I never envied their situation. They rode in the rear seat of the YF-12 with practically no outside view, exposed to some of the "wild things" that we front-seaters put them through. I salute both of these gentlemen for their contributions and their friendship. These two are classic examples of the high quality of NASA people.

The Air Force crews also had backseaters, but they were called Fire Control Officers (FCO). They were responsible for operating the defensive and offensive systems of the Mach 3 interceptor. They YF-12A was equipped with an advanced radar system, and missile launch bays in the forward fuselage. The

FCOs included Maj. Billy Curtis, Maj. Gary Heidelbaugh, and Maj. Sammel Ursini. I had the pleasure of flying with Sam on my first three checkout flights in the YF-12A.

NASA contracted Lockheed, designer and manufacturer of the Blackbirds, to support the program. Lockheed, in turn, had other sub-contractors assisting them with engines, flight controls and other systems.

Blackbird Design

The Blackbirds were highly advanced for their time. A Central Intelligence Agency (CIA) requirement for a high-speed, high-altitude reconnaissance aircraft resulted in an airplane unlike any other. The Blackbirds remained the fastest air-breathing aircraft in the world from the debut of the A-12 in 1962 to the final flight of the SR-71 in 1999. The brainchild behind the A-12, YF-12A, and SR-71 was Lockheed chief engineer Clarence "Kelly" Johnson and his team. Johnson's secret workshop was nicknamed the "Skunk Works" because of the strange and interesting things they "brewed up." The Blackbirds were designed in the late 1950s and made their first flights in secret during the early 1960s. They looked like something from the future, sleek and sharp with blended contours. Even when they just sat still on the ramp, they radiated an impression of speed. The CIA operated the A-12 secretly until 1968. By then, the public was familiar with its sister ships, the YF-12A and SR-71. Those two airplanes, operated by the Air Force, had been revealed to the public in 1964. The SR-71, a spy plane like the A-12, had a long and distinguished career in operational use and as a NASA research platform.

Seeing the YF-12A for the first time, I had the same reaction as when I first encountered the XB-70. I thought, "Wow, what a unique, tremendous aircraft." Its design was startling. The fuselage was long and tapered to a sharp point. The nose radome housed a large radar dish. The thin delta wings had rounded tips and complex curves along the leading edge. Two massive turbojet engines with sharply pointed inlet cones dominated the center of each wing. The two vertical tail fins, mounted on the aft portion of the engine nacelles, canted inward. Looking closely at the leading and trailing edges of the wing, I could make out a pattern. It looked like a series of interlocking triangles. I learned later that this structure helped attenuate incoming enemy radar waves. It was a precursor to what we now call "stealth" antiradar technology.

The Blackbird had been designed from the ground up. All of the technology was new. Lockheed had to learn how to build an airplane almost entirely out of titanium. The airplanes used new types of oil and fuel, designed for very high temperatures. The fuel was even used to absorb heat from the engine oil and ship's hydraulic systems. After taking on this heat, the fuel was then directed to the engines. Two Pratt & Whitney J58 afterburning turbojet engines provided over 30,000 pounds of thrust each to power the aircraft. This was comparable to the XB-70 engines. During Mach 3 cruise, inlet air bypassed the compressor and went straight to the afterburner. Thus, the inlets provided over 80 percent of the total thrust at design cruise speed.

The configuration of the Blackbird required a unique air inlet for each engine. By using a moving spike (with 26-inch travel) located in the inlet, the shock waves were continually positioned within the inlet to provide maximum pressure recovery for the engines. In comparison, the XB-70 used one large inlet to serve three engines. A movable ramp controlled the shock wave by changing the size of the inlet. Thus, two different approaches with the same end in mind.

The Blackbirds had a unique problem with leaking fuel. The structure of the aircraft had built-in fuel tanks. This is called a "wet wing" approach. The tanks were sealed with a gray rubber-like substance that had to be heat-cured after it was applied. Because heat causes metal to expand, designers left slight gaps between the airplane's skin panels and attach points that could slide along the supporting structure. Thus, when aerodynamic heating expanded the panels, they did not buckle and shed in flight. This meant that the airplane leaked like a sieve when it sat on the ground. As chase pilot in the F-104, I was completely amazed one day when Fitz taxied the Blackbird forward out of the parking spot, leaving a silhouette of fuel on the concrete where it had sat.

The special fuel, called JP-7, dripped from every panel line on the bottom of the airplane. When the Blackbird was in the hangar, the ground crew placed buckets and pans underneath to catch the leaking fluids. It was hard to convince the fire department that this was not a hazard. In fact, a lit match dropped into the fuel would be extinguished. Because of the fuel's high flashpoint, it had to be ignited with a chemical called triethylborane (TEB) during engine start.

Starting the aircraft's engines required two large Buick V-8 motors in "start carts." They were placed under the engine nacelles and connected to a mechanical drive that plugged into the bottom of the engine. The Buicks roared as they turned the engine compressor sections at about 14 percent of the required revolutions per minute (rpm) for engine ground operation. When the J58 was up to speed, fuel and TEB were introduced. A green flash of TEB flame exited the aircraft's tailpipe as the engine lit. The start carts continued to assist the engines to about 45 percent rpm before they dropped offline, cracking and backfiring as they finished their work. It was very noisy, but pleasant to hear as it meant the flight was about to begin. It wasn't unusual for people not directly involved with the operation to watch the start up of the Blackbird. We often had a crowd of spectators during preflight, taxi, and takeoff.

We usually carried enough TEB onboard for 16 aircraft starts, afterburner lights, or combinations thereof. There were small mechanical counters by the throttles that were set at "12" prior to flight to leave a little cushion. Normally that was more than enough for a normal flight in the Blackbird, even with air-to-air refueling. However, on certain test flights, we had to limit the number of test points requiring frequent lighting of the afterburners.

Lockheed built the Blackbird primarily of titanium, a metal almost as tough as steel and almost as light as aluminum. It had heat-strength characteristics similar to steel, but its light weight insured that the airplane could carry a usable payload to altitude. In comparison, the XB-70 was made primarily of stainless steel alloys. To reduce the weight, the XB-70 used thin sheets of steel with a honeycomb center structure. The Blackbirds were flexible and structurally limited to 2.5g. On the SR-71, many external structures, such as the chines, tails, inlet cones, and parts of the wing edges, were made from plastic laminates. This added to the stealthy characteristics of the airplane.

Titanium is normally a silver-gray color, like aluminum, but the aircraft were painted with a special "iron ball" black paint that not only radiated heat, but also provided additional radar absorption by dissipating electromagnetic energy to reduce the radar signature. Overall the aircraft was about 107 feet long and had a 55-foot wingspan. The two large engines, inlets and nacelles were located almost at mid wing. Flying in turbulence quickly revealed the flexibility of these wings, as the engine nacelles and large vertical tails twisted and flexed visibly. Each main landing gear consisted of three wheels, with two smaller wheels on the nosegear. The wheel wells were insulated to prevent excessive inflight heating and the main gear tires had a silvery coat of aluminum for additional protection.

The YF-12A and the SR-71 differed significantly in appearance. The YF-12A had a large retractable ventral fin located on the centerline of the fuselage at the rear of the aircraft. The purpose of the ventral was to provide additional directional stability at high Mach numbers. The YF-12A also had two small ventral strakes located under the engine nacelles. The fuselage chine on the YF-12A ended as it approached the radome, resulting in a step or square corner on the chine front. The SR-71s had the very smooth "duckbill" chine that ran completely to the nose of the aircraft. The nose chine actually provided the additional directional stability at higher Mach numbers. The SR-71s lacked the ventral fin, and the strakes only appeared on the SR-71B and SR-71C trainers.

Checkout

My XB-70 experience helped prepare me for checkout in the YF-12A. Both were delta-wing, multi-engine jet aircraft. Additionally, I had become familiar with the use of pressure suits in the XB-70. Similar suits were required for high-altitude operations in the YF-12A. Mainly, I had to learn the Blackbird's systems and their functions.

After many weeks of systems study and briefings with Air Force pilots like Bill Campbell, I was ready to fly the Blackbird. I first made a number of ground engine runs for cockpit, systems, and airplane familiarization. There was only one pilot seat in the Blackbird. The rear seat was for the fire control officer (FCO) in the YF-12A or reconnaissance systems operator (RSO) in the

SR-71. On NASA research missions, the flight test engineer (FTE) occupied this position. In addition to assisting the pilot with systems operations as the FCO/RSO would do, the FTE had to monitor and operate various flight test equipment and instrumentation. Since the Blackbird trainer, located at Beale AFB, was unavailable at the time, I made my first flight in Ship 935 on 1 April 1970. It was the first Blackbird flight with a NASA pilot and it took place on April Fools Day.

Maj. Sam Ursini, an experienced Air Force RSO accompanied me on my checkout flights. Sam, like all of the other Air Force personnel assigned to the project, was sharp and knew the aircraft well. This was bonus for me. I was flying with a living encyclopedia of the aircraft and its systems. Sam monitored systems, backed me up, and gave me guidance throughout the flight. The only thing Sam didn't have in the rear seat was a set of flight controls. I always gave the backseater a lot of credit. In essence, he was saying: "I'm putting my life in your hands, let's go." Sam flew with me on my first three flights and we covered the range of the aircraft's operational envelope to Mach 3.2. Everything went well except on one of our high-speed flights where I experienced a series of violent "unstarts." More on that later.

YF-12 Flight Operations

We conducted intense and thorough flight planning sessions and briefings before each flight. It was standard procedure to have a set of "back up" test cards to be conducted at lower altitude and slower speeds in the event that some malfunction prevented us from flying the high-speed flight profile. This gave us a fallback position and allowed us to gather some data, even on a flight that could not get to high speed. Actually, we didn't have to use the back-up cards very often, and usually any problems we encountered in flight required that we return for landing without delay.

Prior to a flight, the research engineers, pilots, and flight test engineers met to discuss the mission plan. The research engineers presented a briefing on the type of data they wanted to gather. If the data requested was reasonable and achievable, the flight crew suggested what type of maneuver should be flown to generate the required data. The initial flight test cards were already printed up prior to this Technical Briefing, or Tech Brief. The cards presented the test maneuvers required and their sequence during the flight.

The Tech Brief normally took place a minimum of three days before the flight to allow all concerned parties to study the test points and request any necessary changes prior to the flight. As pilot, I made a comprehensive review of the test cards with all of the planned data points. A simulator would have been helpful to practice the test flight, but there was no Blackbird simulator available at NASA FRC at that time. The flight test cards were inserted in a checklist-type book that I attached to the right leg of my pressure suit with Velcro material. The Blackbirds had their own checklists, one for normal operations and one for emergencies. The pilot had to keep all this material available in the cockpit. The aircraft's normal checklist was attached to the top left thigh of my pressure suit. I continually referred to this material during the flight to make sure that the normal aircraft functions were completed along with the test points. The aircraft emergency checklist was placed in a side console pocket for reference, if needed. In an emergency situation, the test card and normal checklist were put aside in favor of the emergency checklist.

Because of the lack of a simulator and the fact that each flight was busy and time-critical, I had designed a substitute for the simulator at my desk at home. I had pictures of the cockpit control panels set up on my desk as they were in the aircraft. Prior to each flight, I sat at this desk, or pseudo-simulator, and mentally ran through the aircraft's normal checklist and the test cards while reaching for and visually noting the positions of each switch that would be activated during the flight. Depending on the complexity of the flight, I went through this practice a number of times until I was familiar with the flight procedures and maneuver sequences.

On the morning of the flight, we had our final briefing. This was called the Crew Briefing, and included any last-minute changes in the test cards. The pilots tried to encourage everyone to keep their last-minute changes to a minimum because any late changes could disrupt prior planning. The Crew Brief also included current weather forecasts, tanker information, and a review of emergency procedures. We also discussed alternate landing fields for use in the event of a problem that prevented us from returning to Edwards. After the Crew Brief, the flight crew proceeded to the life support van to "suit up" for the flight. The pilot and flight test engineer donned their pressure suits, which were just like astronauts' space suits. Even though it was bulky, cumbersome, and reduced the pilot's visibility, it was comforting to wear the suit while flying in the hostile environment of high-altitude. The suits were necessary for our survival if we had to eject, and they had saved the lives of some Blackbird crewmen in the past.

As the life support van transported us to the parked Blackbird, the personal equipment technicians helped us into the pressure suits and ran a pressure check to make sure there were no leaks or malfunctions. When the aircraft crew chief reported ready for pilot entry, the personal equipment men (two per crewmember) accompanied us to the aircraft and assisted us into the cockpit. They hooked us up to the radio, oxygen, and ejection seat. They then performed another oxygen-flow check and suit-pressure check. When the "suit people" left, we were completely tied in, hooked up, and ready to go.

Next, we began the pre-start checklist, which was rather long and involved. Once that was complete and the ground crew was ready, we started the engines. The start carts soon had each turbojet engine turning. The unmuffled V-8s shook the concrete, the turbines began to crank, and a squirt of TEB preceded a bright green flash in the exhaust as the fuel ignited.

Triple-Sonic Blackbird

NASA photo EC76-05086

In-flight view of the YF-12A shows the distinctive ventral fin and sleek profile. Two pods mounted beneath the engine nacelles carried cameras.

Following engine start, we performed another series of post-start checks on the aircraft systems. We had to verify that the flight controls, autopilot, and stability augmentation system (SAS) were functioning properly. About 40 minutes from the time we entered the cockpit, we were ready to taxi. The Blackbird, unlike the XB-70, had an adequate nose-steering system. Visibility for taxi was good and the aircraft's size made it easy to handle on the ground.

We normally conducted our test flights early in the morning, for several reasons. Outside air temperature during takeoff was a critical factor, especially in the summer. We took off with full fuel, which meant a heavy aircraft. Higher runway temperatures increased our takeoff roll. Fortunately, the Edwards runway is nearly three miles long. We preferred, however, to take off with the ambient temperatures below 70 degrees because that provided a margin of safety in case of engine malfunction.

I lined up for takeoff squarely on runway centerline. With the pre-takeoff checklist completed, I increased engine thrust to "military power" (standard cruise power). As soon as I was assured that I had proper thrust, I released the brakes. Almost immediately, I lifted both throttles slightly and advanced to minimum afterburner. The afterburners always lit very close together, but

177

seldom simultaneously. As each one lit, the nose of the aircraft swung in the opposite direction, even with the nose-steering engaged. Once both afterburners were lit, I immediately advanced the throttles to maximum afterburner for takeoff. With a full load of fuel, the aircraft accelerated at a moderate rate. I rotated the nose at about 175 knots and held it about ten degrees above the horizon. The Blackbird became airborne around 210 knots. I retracted the landing gear and reduced engine thrust to military power for a climb at 250 knots.

Climbing through 10,000 feet, I accelerated to about 400 knots. I set up a Mach 0.9 climb to about 22,000 feet and selected afterburners again. The climb continued to about 34,000 feet where I applied a gentle nose-down stick input to produce a 3,500-foot-per-minute rate of descent. This descending acceleration kept the aircraft's angle-of-attack as low as possible during transition through a critical, high-drag region of the flight envelope. The aircraft continued to accelerate to 450 knots equivalent airspeed (KEAS), or about Mach 1.2 at this altitude. Having established a constant 450 KEAS velocity, I began a climb to cruise altitude. When the Blackbird reached an altitude of about 75,000 feet, I could accelerate to Mach 3.2, the aircraft's design cruise speed. The actual maximum Mach number we could sustain depended on the outside air temperature. Parts of the airframe could reach 1,000 degrees from aerodynamic heating, as well as radiant heat from the engines. Too much heat can weaken the structure, even when it is made of titanium. Consequently, we had to pay close attention to atmospheric temperature and Mach number, so as not to exceed the "heat barrier."

It usually took between 20 and 30 minutes from takeoff to cruise altitude, depending on the atmospheric temperatures. The Blackbird, like all jet aircraft, operated better at altitude on very cold days. Acceleration went much faster and the engines consumed less fuel. On a "hot" day at altitude, it took the old girl a lot longer to get up and go. Even before leveling off at the test conditions, I reviewed the first test point and completed any necessary preparations.

The first leg of the mission took us north, toward the Canadian border, and the second leg headed south for a rendezvous with an aerial tanker near Tonopah, Nevada. Heading southbound, we gathered as many test points as possible. The start of our descent was critical in that we had to be on altitude and in a certain location prior to meeting our tanker. The descent and slowdown to the tanker covered about 250 miles, and we were very busy working on clearances with the Federal Aviation Administration (FAA). We had to descend through a major east-west jet corridor and coordinate the crossing with other traffic. Once we dropped below about 25,000 feet, we were clear of civilian traffic and we entered the Air Force test area that was void of commercial aircraft. We operated on a see-and-be-seen basis, but Edwards Radar Approach Control (RAPCON) provided advisories.

I experienced many aerial refuelings during the program. I found it a demanding task that required a great deal of concentration. It usually took about 12 to 14 minutes to top off the Blackbird's fuel tanks, and I was always ready to go back into the climb-accelerate mode and on to other things when that time was up. The YF-12 had poor visibility for refueling. The pilot's seat would not lower far enough to allow the pilot to see the tanker by looking upward. Therefore, he had to bend his torso forward and arch his head back to view the KC-135. It was no easy task while wearing a pressure suit helmet inside a cramped, triangular canopy. Flight control response and damping were satisfactory at lower speeds, especially at lighter gross weights. When the fuel load was between half and full, a structural characteristic manifested itself.

If I made too large a pitch input, I felt a small but discernible bending wave pass through the fuselage structure. Like the XB-70, the YF-12 was a flexible aircraft. The pilot had to be careful not to fight this phenomenon with large stick inputs, but instead reduce his inputs and allow it to dampen naturally. While "on the boom" taking fuel, I occasionally felt this structural wave begin and then dampen. Any air turbulence aggravated the effect. The aircraft's inertia increased with the increasing fuel load and this decreased the aircraft's response to throttle movement. A weight change of 55,000 to 60,000 pounds was not unusual during refueling.

Whenever I flew the Blackbird, Fitz served as primary chase in the F-104 and vice versa. Early in the program, NASA management suggested that the pilot not flying the Blackbird that day should be in the control room, but Fitz and I both believed that we could support each other better in the air. The primary chase rendezvoused with the Blackbird on takeoff. During the initial climb, he looked the Blackbird over to make sure that all gear doors were closed and all other portions of the aircraft looking normal. As the YF-12 began its acceleration through Mach 1, the F-104 chase slowly dropped behind to conserve fuel, keeping the Blackbird in sight by watching the contrail behind it. If a problem developed, the F-104 rejoined the Blackbird and escorted it home. As the YF-12 climbed higher, it exceeded the altitude where contrails formed and the chase pilots called the control room on the radio to say that they had lost visual contact with the Blackbird. The YF-12 pilot then called the control room with his altitude to mark the level at which the contrails disappeared. This was usually around 60,000 to 65,000 feet (about 12 miles above the Earth). From this point, the F-104 pilots set their engine power for maximum endurance and held position just west of Beatty, Nevada, in the refueling area.

We preferred to use the tanker support after the first long leg of the flight. In that way, we were able to "get two full flights for the price of one." The onboard data recorders on the NASA Blackbirds were good for a little over two hours of data, and if we put our tanker in the middle of the mission, it permitted us to fill our data tapes. Once we spotted the Air Force KC-135Q tanker entering the refueling track, the F-104 chase pilots made radio contact and joined up with the tanker in a loose formation. This allowed the chase to watch for the returning Blackbird, coming southbound into the refueling track. There were occasions when the chase assisted the YF-12 pilot by calling out the tanker's position and range. Both Fitz and I appreciated this because a little moisture on the pressure suit visor, formed during descent, sometimes limited long-range visibility from the cockpit. As the YF-12 joined in formation with the tanker, the chase crews took the opportunity to look the YF-12 over for general integrity and make sure that there were no fluids leaking from the aircraft. After the Blackbird received its fuel and broke away from the tanker, the chase pilot gave it one final look, checked that the refueling door was closed, and wished the pilot a good second leg. By this point, the F-104s were low on fuel and ready to return to Edwards. Sometimes the primary chase pilot parked his empty F-104 and immediately climbed into another aircraft that was waiting for him. He then took off to greet the Blackbird as it entered the Edwards area for landing after its second flight leg.

Once off the tanker and cleared to climb, we lit the afterburners and started our climb acceleration from the tanker altitude of around 22,000 feet. This test leg was usually a little longer than the first because we did not have to burn fuel during taxi, takeoff, and initial climb. I accelerated back to speed and altitude going north toward Canada. Normally, I could complete several test points before turning south to avoid crossing the border. That sometimes required a 70-mile turn at Mach 3.2. During the turn, I finally had a little time to look around outside the aircraft. Up to that point, I had been busy establishing test conditions and taking data, and flying mainly on instruments. The actual outside horizon, visible through the cockpit windows, served as a backup to the artificial horizon (attitude indicator) on the instrument panel. On lower, slower-flying airplanes, the pilot could do a pretty good job of flying just with the visual reference to the outside world. In the Blackbird, the speed was too great and fractions of a degree in pitch resulted in large altitude excursions. This aircraft was operated primarily on instruments regardless of weather conditions.

The aircraft had a good autopilot, and I used it whenever possible to relieve pilot workload. I usually made turns on autopilot, but the majority of the flight was flown manually due to the requirements of the various test points. The view outside was spectacular, especially if the weather was nice and there was not too much cloud cover. It was amazing just how far I could see. The slight curvature of the earth at high-altitude and the beautiful shades of blue in the sky were unlike anything I had seen from other airplanes. The sky color ran from a light to a very dark blue, almost black, as I looked upward. As the turn came to an end, it was time to get back to work on the test points.

The cockpit on the Blackbird was designed for a pilot in a full pressure suit. It was possible to reach all of the switches and buttons with little effort, with the exception of those on the consoles alongside and slightly behind the pilot. The pressure suit made it difficult to reach these, but it was possible with some effort. They were not the most critical items, anyway. Some of the major flight instruments were of the tape, or vertical-strip, variety. I had experienced these in the XB-70, and although I preferred the round dials, I had no problem flying with the tapes.

Visibility outside the aircraft was good, looking forward and up, but it was limited looking back or down. There was a small periscope in the top of the canopy that could be extended to allow the pilot to check his engine nacelles and vertical stabilizers. It was difficult to use, but again, it was possible with some effort. The periscope was normally retracted at high Mach numbers, but I forgot about it on several occasions and it didn't slow the aircraft down a bit. It was only an inch in diameter and extended only an inch or two outward from the fuselage. The forward canopy was angular and had thick glass and a thin dark splitter plate to prevent reflections. I thought that this combination along with the pressure suit helmet visor might give problems with visibility during takeoff and landings, but I was wrong. There was no problem. I was never sure if my left eye looked out the left of the splitter plate and my right eye looked out the right, but I had no problem with judging height during landings.

The pilot scans numerous instruments in the cockpit in order to assimilate and react to the information presented. This information relates to what the aircraft is doing with changes in attitude, altitude, and velocity. In addition, he must monitor the performance of all the aircraft systems, including flight controls, fuel, hydraulic, electrical, and propulsion systems. It seems like almost an impossible task, but using the "pilot's scan" technique establishes a set of priorities for reading the instruments.

Using the scan technique, the pilot visually scans the instruments in a regular pattern and digests the information. He repeats the scan pattern, providing his mind with updated information on each pass. The way the pilot prioritizes this is to give the most critical items, such as aircraft attitude and speed, a first priority, aircraft propulsion a second priority, and other aircraft systems such as fuel, and electrical a third priority, and so on. In addition to the instruments, there are a series of warning lights to catch the pilot's eye in the event a system exceeded a normal limit on the instrument while the pilot was scanning elsewhere.

A warning light immediately changes the pilot's scan priority, and he immediately has to give attention to the instrument related to the warning light and determine the required action. Of course, the pilot must be thoroughly trained on the appropriate response for an out-of-limits condition. At Mach 3 speeds, it is not possible to break out the handbook for reference. The pilot has an abbreviated checklist that he may refer to for help. The scan is continuous as is the pilot's response to it in controlling the aircraft by making inputs to the flight controls, throttles, and other systems switches. On autopilot, the pilot's task is reduced to monitoring the aircraft's position and attitude, and fine-tuning the autopilot with small corrective inputs. During climb acceleration, I was able to use the autopilot to great advantage. During the test points, the aircraft was hand-flown because of the unique inputs required for collecting data points.

The YF-12 had fine low-speed handling qualities, and approach and landing speeds similar to the F-104. Control response in the landing configuration was excellent and stability was very good. The Blackbird had no angle-of-attack indicator in the cockpit. The airspeed indicator served as the primary approach reference. It was easy to set a desired airspeed and control the glide path with the throttle. Landing flare and touchdown were similar to other delta-wing aircraft. It had good ground-effect and feel, and smooth landings were normal. The high-speed flight characteristics were good, as that was the flight regime the airplane was designed for. The control systems design provided the pilot with good feel and response over the complete flight range. It was an excellent airplane.

Unstarts

On my second flight to high Mach number in ship 936, I experienced a series of unstarts and restarts that rattled my head and other parts. The fact that the Blackbird could operate at speeds from low sub-sonic to Mach 3.2 is impressive. The systems that made this possible included a computer that automatically positioned the inlet spike to control the shock wave's location, and bypass doors to control the air pressure and flow in the inlet. Both of these are critical to provide the engine with "prepared" air with a high static-pressure and a subsonic Mach number. The pilot has manual controls as well, but it requires a great deal of the pilot's time and attention. The engine inlets and controls function very much like a wind tunnel in front of the engine. At supersonic speeds, when the shock wave is located inside the inlet, everything runs smoothly. If a situation develops that upsets the inlet flow to a point that the position of the normal shock wave moves out of its optimum position, the interrupted airflow results in an unstart.

The unstart condition is not a gradual thing. It happens violently, with a loud bang. When the inlet unstarts and loses its smooth flow, the aircraft tends to pitch-up. At the very least, it can snap the pilot's head and helmet against the inside of the canopy. At worst, it can snap the aircraft in half. Fortunately, the flight control stabilization systems have a great deal of power to counter the initial directional divergence that accompanies an unstart. In addition, a special "cross-tie" feature is included in the inlet controls to automatically unstart the other inlet to reduce aircraft directional divergence. The cross-tie is active above Mach 2.3, and even with this feature, unstarts can only be described as violent under certain flight conditions. In other supersonic flight conditions, they are less violent, but still disconcerting. As a comparison, an inlet unstart in the XB-70 was less violent than in the YF-12. On the XB-70, in fact, a failure of one engine out of the three might not even unstart the inlet. It just altered the position of the throat and bypass doors. The XB-70's engines were also located near the centerline as compared to the Blackbird's mid-wing location.

Blackbird unstarts were unique and, at times, a little scary. The inlet control system had a feature that was called auto restart. When the inlet unstarted, the computer control automatically retracted the spikes and opened the bypass doors to a position associated with a lower Mach number. The spike then moved forward slowly and the bypass doors closed down as the shock waves repositioned. If the original cause of the unstart was gone, the inlet restarted and airflow was restored to the engine. In general, the Pratt & Whitney engines were quite forgiving, and they included a special feature in the fuel control to reduce fuel flow to the engines during unstart events. The natural tendency of the engine during a reduction in airflow during unstart was to exceed the design temperature limit and cause damage to the engine. This feature was called "de-rich."

On one occasion, the aircraft went through the automatic restart sequence, but the problem that caused the unstart was still present. The inlets unstarted again. After several cycles, I selected a manual restart position and held the inlets in this "safe" condition while the aircraft slowed. About this time, poor old Sam Ursini, who couldn't hold it back any longer, said, "Mallick, what in the hell is going on up there?" I calmly said, "Sam, as soon as I find out, I'll tell you." As it turned out, I released the inlets from manual and the next restart worked. We were back in business. One of the most disconcerting things about inlet unstarts was that after a flight, ground checks sometimes revealed nothing wrong with the inlet controls or computers. The problem was dismissed with the belief that the computer was setting the inlet position too close to the unstart boundary.

On one exciting ride, I was close to Edwards before I reached my 30-minute heat-soak time at Mach 3. I decided, however, that I had enough airspace to complete the maneuver and still slow down for landing at Edwards. As I pulled the aircraft's nose up, the inlets unstarted and the engines went into a partial stall condition. Cabin pressure dropped to a point where our pressure suits began inflating. The nose of the aircraft continued upward for a period much too long to be comfortable to me, considering I had immediately applied full-forward stick without apparent effect. As the nose finally started to drop, the inlets automatically attempted to restart. This failed. After several more attempts, the inlets restarted, the engines cleared their stall, and I was able to commence a turn back toward Edwards.

Unfortunately, as all of this was going on, I was heading out of the Edwards test range toward the Los Angeles Control area, which had many airliners operating at any given time of day. We always had an FAA representative in our control room during the flights. He was in contact with the various FAA centers that controlled the airspace we operated in and around. It really paid off on this mission. The FAA Coordinator integrated my intrusion into Los Angeles Center airspace and indicated that I should stay above 45,000 feet until I returned to the Edwards airspace if possible. It was possible, and that alleviated the situation. I turned back toward Edwards from over Pasadena, still humming along at 50,000 feet altitude and well over Mach 2. After landing, I went over the flight control data traces with the engineers to determine how long I had held full-forward stick until the aircraft nose attitude had started down. It had only been about eight to ten seconds. With both inlets unstarted, the pressure suit inflated and there was a "whole lot of shakin' going on," and it had seemed like ten minutes to me at the time. For a moment or two, I wasn't really sure the aircraft was responding to my input and I was concerned about who was in control.

Several hours after landing and debriefing at NASA, I learned of a phone call that indicated that our flight had not escaped all notice that day. As I completed the 180-degree turn back toward Edwards, our shock wave had broken a large plate glass window in a bar in Pasadena. Fortunately, nobody had been injured by flying glass, and NASA picked up the repair check because we knew the abnormal flight path projected a pretty strong sonic boom onto the ground. We were happy that there was only one broken window, or perhaps just one reported.

Aerial Refueling

Aerial refueling was a complex task, but necessary during long duration flights. We typically rendezvoused with the tanker at about 20,000 to 22,000 feet altitude in an area called a "refueling track." Here the tanker flew an oval racetrack pattern as we climbed up from below and behind. A crewman, called a "boomer," lowered a refueling boom at the rear of the KC-135. The boom had two control vanes and an extendable nozzle. The boomer maneuvered the control vanes as we slid into position below the tanker. He then extended the nozzle until it locked into the Blackbird's refueling receptacle on top of the fuselage. Once the nozzle was locked into place, the fuel flowed from the KC-135 into our waiting tanks. I always had to pay close attention to my position relative to the tanker. I had to make constant adjustments to my power and control settings. When the Blackbird's tanks were almost full, the boomer adjusted the number of transfer pumps to allow the maximum transfer of fuel before pressure sensors automatically cut off the fuel flow.

We made contingency plans, in case we were unable to take on fuel for some reason. We always had another airfield in range that we could divert to, if necessary. On one occasion, while refueling near Beatty, Nevada, I was unable to stay connected to the tanker. I hooked up, got a green light, and immediately the Blackbird fuel receptacle spit the nozzle out. After trying numerous times, I decided to return to Edwards. A mechanic later found a fault in the electrical portion of the Blackbird's refueling system. On that flight, we just couldn't complete the second leg of the test mission and we had to abort.

Turbulence during refueling added greatly to everyone's workload. The tanker normally flew on autopilot because it provided the most stable platform during refueling. If the tanker was flown manually, crew workload increased, even in smooth air. A good autopilot is inherently smoother on the controls than a good pilot. A human is not as quick to sense changes or, to put it another way, the electron is quicker than the eye. One of the most difficult refuelings I ever experienced was on a special mission in which we decided to fly out of our normal north-south test area. Instead, we operated further east, in a refueling track near El Paso, Texas. As it turned out, Edwards was too far away for an abort and our contingency airfield was not very desirable for landing. I was concerned about the fact that I did not know the surrounding ground references in the refueling area. At Beatty, I knew the area so well that I could tell if the tanker was a little off track when we met up. The YF-12A did not have the sophisticated air-to-air rendezvous equipment that the SR-71 had, so we did our rendezvous using relative location to a ground station, ground references, and visual sightings.

In the El Paso rendezvous, this worked out fine and we spotted our tanker in good time. The air, however, was very turbulent. On top of this, the tanker pilot informed me that he planned to fly the rendezvous manually in order to complete some training. It was the toughest refueling that I ever accomplished and I experienced turbulence-induced boom disconnect on several occasions. I was tempted to ask the Air Force pilot to use the autopilot, which would have helped some, but I was too proud for that. I worked my ass off and broke a good sweat, but I got the fuel. After all, I didn't want to land at an outlying base in Southwest Texas because I couldn't refuel. During the postflight debriefing at Edwards, I offered the opinion that we should stay with our standard north-south test area and the Beatty refueling track. From then on, we did just that.

I don't want to leave the impression that all my refuelings were difficult. On some flights, I just pulled up to the tanker, hooked up, and refueled as if it was "old hat." When we were on the boom, I flew with reference to the tanker. There was a series of director lights on the belly of the KC-135 that showed me where I needed to be. On an easy day, I sat right "in the green" (the center of the director lights) for the whole refueling, and only started to drop back when the aircraft reached full load and the power was set at full military thrust. When the air was smooth, the tanker stable, and I had a recent refueling under my belt, I was in top form. Unfortunately, we didn't average a flight per crew every two weeks. It was usually longer between flights, sometimes months.

After dropping off the tanker with a full load of fuel, it was time to quickly catch my breath while setting the Blackbird up for the next climb and acceleration to test conditions. The very first requirement after having the F-104 chase look us over, was the clearance from the FAA center to start our uninterrupted climb to the north. In order to "grease the skids," we had conducted FAA briefings describing our mission and the importance of time and fuel. We traveled to the FAA centers that were involved and briefed as many of the controllers as were available. The FAA was generally very cooperative and helped us get a higher priority for our clearances. If a conflict arose, the controllers vectored commercial traffic around our path in order to allow us to proceed on our mission.

There were times, of course, where it was impossible for the controllers to do anything but hold or divert us, and we followed their direction because they had the primary responsibility of keeping aircraft separated in their areas. It worked out very well. Several years after our initial briefing to the FAA, we could tell that some of the "briefed" controllers had moved on to new assignments. We briefed the new controllers because we knew from experience that it helped a great deal and made our test flights more productive.

Loss of a Blackbird

The two airplanes had been flying for about a year when 936 was lost in a crash on the east shore of Rogers Dry Lake. The Air Force crew, consisting of Jack Layton and Billy Curtis, was on approach to Edwards following a test mission to collect handling qualities data. An intense fire broke out in the right-hand engine. A fuel line had broken and the situation deteriorated rapidly. Jack attempted to reach the lakebed for an emergency landing, but decided it was too dangerous. Both crewmen ejected safely despite the low altitude. Ship 936 pointed its nose toward the earth and blasted a deep crater in the desert. There was little left to recover.

By this time, NASA had reached an agreement with the Air Force to acquire an SR-71. There had been some political complications, due to the sensitive nature of the aircraft. The Air Force agreed to loan the second production SR-71A (serial number 61-7951) to NASA, but without some of the improvements that had been incorporated into operational aircraft. The airplane had the same type of engines as the YF-12A and lacked the plastic inlet cones of the newer SR-71s. Additionally, in NASA service, it carried a bogus designation and tail number. SR-71A (17951) became YF-12C (06937). Thus, the public saw it as simply another version of the YF-12, and the tail number was in sequence with the other YF-12A tail numbers. Preparations to add Ship 937 to the fleet began even before the crash of 936.

The Air Force program was nearly complete and NASA crews continued to fly the airplanes in support of radar tracking and intercept missions for a while. We usually incorporated these intercepts into our own research flights. On our last leg south coming down from the Canadian border, we served as a Mach 3 target for Air Force F-106s and F-4s, and Navy F-14s. Ground radar

Lockheed photo LA-4087

YF-12A number 936 was lost in a non-fatal accident on 24 June 1971. The Air Force crew ejected after the right engine caught fire due to a broken fuel line.

vectored these aircraft until they picked us up on their own systems. At this time, they commenced a Mach 5 closure intercept and, at a given distance, they pulled up to over 50 degrees climb angle to time a simulated launch of an air-to-air missile. It was interesting to watch the interceptor contrails as they approached us from the south around 35,000 feet with a closure rate of five times the speed of sound! We were cruising at Mach 3.2, straight and level at nearly 80,000 feet. At the pull-up point, which we could usually see by the contrails bending upward toward us, the interceptors rapidly climbed above the contrail level and disappear from our sight. We then heard "flameout" calls from the interceptors as their engines quit at high altitude and low speed. I estimate that their top altitude on the zoom climbs was somewhere around 60,000 feet. The results of these tests were not available to us because they were classified, and our mission as a target did not require knowledge of the interceptor's success. My observation, however, was that it was a difficult task for the fighters, and their rate of success was not very high. There were times when the ground-controlled intercept (GCI) tracked both the interceptors and the target YF-12 on radar, but the interceptor's radar failed to lock on and we sailed merrily by. There was only one chance per Blackbird flight as we were on our way home, low on fuel and ready to land.

Research Programs

During the early part of the joint research program, the Air Force used the two aircraft for tracking and radar intercept studies. NASA took this time to collect stability and control data and study handling qualities. Most of the Air Force tests were assigned to Ship 936. The loss of 936 heralded the end of the Air Force portion of the test program. Even after the addition of the YF-12C, only NASA crews flew the aircraft for NASA and USAF tests. Meanwhile, NASA technicians at Edwards prepared Ship 935 for a comprehensive structural heating study.

All aircraft are subject to structural load stress on the airframe. At speeds above Mach 2, however, a new factor emerges: heating loads. Aerodynamic heating from skin friction raised airframe temperatures to over 700 degrees Fahrenheit in some locations on the Blackbird. Coupled with heat generated by the engines, the nacelles experienced over 1,000 degrees at some points. Excessive heating can weaken the aircraft structure and lead to catastrophic failure. NASA initiated a program to study both aerodynamic heating and maneuver loads on the YF-12A. It was a perfect platform for such experiments because it was capable of maintaining high-speed thermal loads long enough for temperatures to stabilize during cruise.

Aircraft 935 was completely instrumented with thermocouples, strain gauges, and deflection measurement devices on the fuselage and one wing. Because the loads were assumed to be symmetrical, NASA technicians only instrumented one wing and nacelle. We collected flight data at cruise speeds of Mach 3.2 after "heat soaking" the aircraft for about 30 minutes. For maneuver loads, data was recorded during what we called a "yo-yo" maneuver. This consisted of pulling the nose of the aircraft up gently and, as the nose came above the horizon, starting a gentle pushover to 0.5g until the nose dipped below the horizon. I then initiated a steady 1.5g pull until the nose came back to or slightly above the horizon again. Finally, I stabilized the aircraft in level flight. The goal, using gentle and smooth control inputs, was to keep the Mach number from varying too much from 3.2. The maneuver sounds mild compared to other fighter-type maneuver load test points, but with a maximum g-limit of 2.5 and the possibility of unstarts that could cause the aircraft to diverge, it was just the right technique. Due to fuselage bending, deflections of the airframe between nose and tail exceeded twelve inches over the 107-foot length of the aircraft during these maneuvers.

During test points on the heat-loads program, the aircraft had to be handflown because of the unique inputs required for each data point. We conducted maneuvers at various specific Mach numbers up to Mach 3.2, our design cruising speed. We were especially interested in looking at data from the extreme end of our flight envelope. In order to collect meaningful data, the aircraft temperature had to be stabilized for at least 30 minutes at Mach 3 before we initiated the yo-yo maneuver. We preferred to perform the maneuver relatively close to Edwards for accurate radar and telemetry data. Therefore, we tried to accelerate to Mach 3 as quickly as possible while headed north, after leaving the tanker, and then continue the heat-soak during our turn and as we turned south toward Edwards. We conducted the Mach 3 yo-yo maneuver near Edwards, on a southbound heading. On 23 February 1972, Ray Young and I refueled over El Paso and made a dash back to Edwards that resulted in the aircraft's longest Mach 3.2 heat-soak of the program.

The instrumentation recorded total loads, a combination of maneuver and heating loads. NASA scientists developed a unique way to separate the two types of loads data so that each could be examined by itself. A ground test facility at FRC, called the High-Temperature Loads Laboratory (HTLL), was prepared with a large cocoon-like jig lined with quartz heating lamps. The assembly was designed to conform to the shape of the Blackbird's exterior. The engines were removed, but heaters were placed inside the nacelles to duplicate equivalent engine heat to ensure the accuracy of the experiment. The data tapes recorded during flight tests were used to provide a time-temperature history of the aircraft. Thus, the Blackbird on the ground experienced the same heat profile as it had in flight. At the same time, the strain gauges on the aircraft provided load data that was due specifically to heating, thus distinguishing between the two types of loads.

When Ship 935 entered the HTLL, the flight crews were concerned about its care and handling. We were worried that the ground tests might damage the aircraft, preventing it from conducting further flight operations. Consequently, we monitored all planning meetings to check on safety issues. On the first attempt to run a complete flight profile in the HTLL, smoke started rising from the fuselage of the YF-12A. This alarmed the heat facility test personnel and they quickly shutdown the heating system. A meeting was called and, after study of the problem, we decided that this was not too unusual. The aircraft contained a large amount of insulation material, such as thermal blankets around the landing gear. Because the aircraft always suffered from fuel leaks, we concluded that the smoke came from residual JP-7 that had soaked into the insulation. Because the JP-7 was so resistant to burning, only a few additional fire-fighting units were placed near the aircraft, and the heating tests continued. We figured that the aircraft probably emitted smoke during flights, but we hadn't been able to see it behind us. Since the YF-12A remained in the HTLL for about a year, we began propulsion research with our second airplane.

Propulsion Tests in the YF-12C

The so-called "YF-12C" served mainly as a propulsion systems testbed. We used Ship 937 to collect flight data from the engines and inlets at hundreds of test conditions, many of them "off schedule" (out of the normal envelope). NASA researchers hoped to use the data to build a technology database for development of a supersonic transport (SST). Like the YF-12A, the YF-12C had "Type J" engines with fixed compressor inlet guide vanes and a maximum afterburner thrust of 32,500 pounds at sea level. An Air Force crew delivered that the airplane to FRC on 16 July 1971, and it was grounded nearly a year so technicians could install instrumentation. By the time 935 was placed in the HTLL, the work was nearly finished.

Fitz and Vic took 937 up for a checkout flight on 24 May 1972. Aircraft checkout flights were also helpful to the flight crews because we could gain some flight proficiency without the added pressure of having to gather data at precise test points. Research flights required a great deal of skill and concentration. Also, there were sometimes long gaps between Blackbird flights for any given crew, making it difficult to remain proficient. If a crew had more than a month between flights, it showed in their performance.

Propulsion research flights in 937 encompassed some of the most demanding flying I ever did in the Blackbird program. Prior to installation of the instrumented engines in 937, they had been put through a series of tests in the NASA Lewis Research Center supersonic wind-tunnel. The engines had been run at a variety of simulated speeds up to Mach 3, with various inlet-spike and bypass settings. During these tests, inlet unstarts were induced by setting the spike position and bypass doors out of their normal operating envelope. Now, the flight crews had to match these wind-tunnel test points closely, in actual flight conditions, to within 0.05 Mach.

In an average test flight, a pilot establishes a test point (specific altitude and speed) at a stable condition for a given aircraft configuration. If the Mach number or altitude is off by a small amount, but the condition is stable, then the test point is considered successfully achieved. With the YF-12 propulsion test points, this was not satisfactory. We needed to match the wind-tunnel data points precisely. This was difficult to accomplish with the existing flight control instruments and the information available to the pilot. In order to help the pilot accomplish these tests, technicians set up the Blackbird pilot's instrument panel to display two new flight parameters from the ship's inertial navigation system (INS). These included the aircraft's inertial rate of climb and inertial longitudinal acceleration. Any change in the aircraft's rate of climb or speed was immediately presented to the pilot, enabling him to make quick corrections. Without these two parameters, the crews had little chance of matching the wind-tunnel points. The critical flight parameters were telemetered to the ground control station and, when conditions were matched, controllers advised the flight crew to activate the data recorder.

I used the autopilot frequently during the propulsion study. It held the aircraft steady for me while I took manual control of the aircraft's inlet system. Then, in small steps, I slowly moved the spikes and bypass doors until they

Following takeoff from Edwards AFB, the YF-12C takes on fuel from a KC-135Q tanker. Refueling was a demanding task for both aircraft crews.

exceeded the operating envelope and the inlets unstarted. In the high-performance region of the inlet envelope, the engine and inlet were at their most efficient setting. Sometimes, I had to gently pull back the throttles to maintain a constant aircraft speed. Unfortunately, in this region of the performance envelope, the unstart margin was extremely narrow. As I progressed with my incremental adjustments, the cockpit grew quieter - I knew it was the calm before the storm. Shortly thereafter, an unstart would occur along with all the aircraft motions, noises, and excursions. Even after flying dozens of these test points, the unstart phenomenon was startling, but it signaled that the test point had been accomplished.

After an unstart, I had to immediately monitor the automatic inlet restart system to make sure that it functioned properly. Next, I checked engine conditions to see if either engine had suffered a compressor stall or had de-riched. I also checked the afterburners to see if they had blown out. I could relight them with squirts from the TEB reservoir. The engines had automatic systems that prevented overtemp situations, and stalls generally cleared quickly. The afterburners almost always had to be re-lit because unstarts blew them out. After accomplishing the immediate tasks, I worked to regain the speed and altitude that were lost as a result of the unstart, and prepared for the next test point. It was challenging, but satisfying, to accomplish a number of good test points during a flight.

During the propulsion flights, we generally flew the same track that was used for our heating and loads missions in Ship 935. This route had been carefully selected to maximize coverage by NASA's tracking radar located at Beatty and Ely, Nevada. The stations were part of the High Range tracking array that was built to support the X-15 flights. The High Range provided not only good telemetery reception, but also space positioning data on almost all of the flight. Only at the northern end of our track, near the Canadian border, was the Ely radar unable track us.

As with the heat-loads test missions, the turn around Spokane at the north end of our track gave us a chance to catch our breath. We never conducted any tests during the turn. It merely served to turn us back toward Edwards. I flew the turn in autopilot, giving me time to study my test cards in preparation for next test point. Normally, we began to set up our test conditions as soon as we rolled out southbound. The Blackbird covered ground so quickly that we were always in a hurry to get on with the test points and accomplish as many as possible before reaching the descent point.

On 2 May 1973, Ray and I were both catching a little breather about half way through our northern turn and everything was running smoothly on autopilot. Suddenly, all hell broke loose as our right engine suffered a violent unstart. The automatic restart mode cycled the spikes and inlets, but the engine unstarted again. After a second auto restart failure, I tried a manual restart and attempted to reset my spike and bypass doors. As I released the manual restart switch, I had an unpleasant surprise. The inlets failed to recover. We were losing altitude and energy as we descended, becoming subsonic around 30,000 feet above the ground. Ray notified the FAA Center, which was the only agency that had us on radar at the northern end, and informed them that we had an emergency. They immediately cleared us for flight at that altitude.

I managed to get the aircraft under control. We were still flying, but using up fuel at an alarming rate. Ray and I started a fuel-range planning review and quickly concluded that we could not make it back to Edwards with our remaining fuel. The Blackbird was designed to cruise at Mach 3.2 and, at that speed, it was most efficient in terms of fuel consumption. Now subsonic, our range had decreased substantially. Our nearest contingency landing field was located about a hundred miles southeast of Reno, Nevada, at Naval Air Station Fallon. We told the Navy controllers to expect us and pointed our nose toward the salt flats of north central Nevada.

Our landing was uneventful. The Navy had recovered Blackbirds there before, Air Force SR-71s out of Beale AFB that also used Fallon as a contingency destination. The ramp crew directed us to park directly in front of the Base Operations building. They also assisted Ray and me out of the aircraft, a major chore due to the pressure suits and attachments. We spent the night at NAS Fallon in the Bachelor Officers Quarters (BOQ). The next day, a KC-135Q arrived from Beale with a load of JP-7. Ground crews fueled our aircraft and installed some new tires. Our high-pressure tires had suffered cuts and bruises due to debris on the runway and taxiway. We had also taxied over an aircraft arresting wire, standard equipment at Navy airfields, that was rough on our tires. The Navy personnel were good hosts. They were also aware of the security concerns surrounding the Blackbird. The chief of security approached me with about a dozen cameras that he had confiscated from curious Navy personnel. He asked me what to do with them. I suggested that he keep the film, but return the cameras. I told him to ask the people to just look, but take no pictures. That was what we were doing at Edwards at the time and it seemed appropriate.

Once the airplane was fueled, Ray and I flew back to Edwards at 30,000 feet in the subsonic mode and in a normal flight suit coverall. It was the only time I ever flew the aircraft without a pressure suit, and I wouldn't have had that opportunity if not for the emergency. It was a bit more comfortable flying without the bulky pressure suit and I found I had better visibility.

Blackbird Coldwall Experiment

Having two operating Blackbirds provided NASA with a tremendous opportunity for high-speed research. Following the heating, loads, and propulsion programs, researchers proposed using the airplanes to carry a unique experiment package to study the effects of high-speed turbulent boundary layer and heat transfer. Called the Coldwall, the experiment consisted of a 13-foot-long stainless steel tube, mounted on a ventral pylon beneath the forward fuselage of the YF-12A. The instrumented cylinder was cooled with liquid nitrogen and covered in a special insulating material. Our flight profile called for us to accelerate the airplane to Mach 3 and then pyrotechnically remove the insulation, instantly exposing the super-cooled tube to aerodynamic heating.

The insulation blanket, several inches thick, enclosed an explosive primer cord. The insulating material was supposed to be light enough, and come off in such small pieces, that it would not damage the aircraft or harm anyone on the ground. As a further precaution, the insulation was to be removed over a remote portion of our test track, further reducing the threat to anyone on the ground. It sounded like a good system, but it gave us a great deal of trouble over the course of the program.

We began flying with the Coldwall experiment package in February 1975 to get some baseline data. Because the flight profile called for operations at Mach 3 speeds and altitudes above 70,000 feet, we used the YF-12C as our primary chase aircraft. The F-104 served as chase only during low-altitude, low-speed portions of each flight. The program was plagued with trouble from the beginning. The first few check flights were aborted early because the accelerometer on the Coldwall pod malfunctioned. That was merely annoying. The third flight was a doozy!

On 27 February 1975, the test card called for an aileron pulse with the roll/yaw stability augmentation system (SAS) off to gather sideslip data. The SAS limits the sideslip angle of the aircraft to protect the airframe from structural failure. An aileron step input generates a yaw response from the aircraft. In this case, the yaw created a load on the ventral and its strain gauges. Since there was no telemetery readout from the strain gauges, the results of the test maneuvers were only known after the aircraft landed and engineers had studied the data. Because we conducted very little flight operation in the Blackbirds with the SAS off, I was conservative with my aileron inputs. Data from a previous flight the engineers requested a larger input on the next flight. The test conditions called for flight at Mach 0.95 (high subsonic speed) and 25,000 feet altitude. This resulted in a relatively high "q" (dynamic pressure) on the aircraft and a larger load on the airframe for a given sideslip angle.

Einar Enevoldson flew chase in the F-104 as I established my test conditions and turned off the roll/yaw SAS. I introduced a very rapid lateral input to the stick and the aircraft responded as expected with a roll and accompanying sideslip. Almost immediately, I heard a sharp "crack" toward the rear of the aircraft, followed by Einar's comment: "I say, something just fell off your aircraft." When he moved in closer, he found that the ventral fin was missing. Also, there was liquid streaming out of the lower fuselage of the aircraft and it was difficult for Einar to tell if it was fuel or hydraulic fluid. I immediately turned the aircraft toward Edwards and requested clearance for an emergency landing. Fortunately, the hydraulic pressures remained normal, but we were losing fuel. I made a quick, straight-in approach and landing, turned off the main runway, and shut down the engines. We made our next six flights without a ventral fin.

The Coldwall flights resumed in July with missions to check the structural characteristics of the experiment package. We also made some "Hotwall" flights without the liquid nitrogen coolant or insulation. Fitz and Vic flew the first true Coldwall mission on 21 October 1976. Premature loss of the insulation prevented collection of any useful data. This happened on several other occasions, ruining our data each time. We didn't get any good Coldwall data until 23 June 1977. The experiment cylinder was mounted aft of the inlets to prevent any insulation debris from entering the engines. It didn't always work that way in practice, resulting in some "hair-raising" moments.

I have described the challenges of aerial refueling and flying precision data points, but these tasks pale in comparison to flying two Blackbirds in close formation at Mach 3 and high altitude. Just holding my position in normal cruise was tough. The aircraft had a slow thrust response at those flight conditions. Because the Blackbirds have no speed brakes, I had to add or subtract power, as necessary to maintain my position when I flew chase in the YF-12C. Flying formation with the reduced visibility of a pressure suit helmet, and very narrow cockpit windows, only added to the chore. Another problem with formation flying at Mach 3 was an extremely powerful shock wave streaming off both aircraft. As a result, I had to position my airplane further forward than where I normally flew for good visibility of the other aircraft. I had to control the airplane while looking out almost 90 degrees to my right, through the tiny side window in the canopy. Under the best conditions, this was bad. On 21 July 1977, it got much worse.

Fitz and Vic were flying the YF-12A and I had chase duty in the YF-12C with Ray. When I was in what I thought was a good position, I called Fitz on the radio and told him I was ready. Fitz gave a countdown and activated the switch that fired the primer cord. Insulation from the Coldwall tube disintegrated in a white flash, but some of it was ingested into the left engine of the YF-12A. Mission planners had told us that this would never happen. The Mach 3.0 airflow would carry all of the explosive gas and debris back behind the aircraft. Apparently, this was not the case. Some debris moved forward and passed through the left engine, resulting in an unstart! The right engine also unstarted, and Fitz had his hands full.

As I struggled to stay in position, I experienced a series of unstarts, too. My aircraft protested all of this off-schedule inlet operation, raising hell with thumps, bangs, and afterburner blowouts. Fitz, in the meantime, finally restarted his inlets. We finally got together again and limped home to Edwards. We landed safely, but both airplanes were grounded until September for inspection and repair. All of the hammering and abuse resulted in a number of cracks in the titanium inlets on both aircraft. After the airplanes returned to flight status, we collected some good data from the remaining Coldwall flights. Data from the experiment eventually validated a leading heat transfer theory.

I decided against further formation flying with the two Blackbirds. It placed both aircraft at risk of a possible mid-air collision. I didn't believe we really needed a chase after the first Coldwall test. We confirmed that the YF-12 crew could successfully remove the insulation and survive. The airplane just slowed down a little in the process. On the first formation flight, some of the research people asked me to gather more propulsion data by flying the YF-12C through the wake of the YF-12A while cruising at Mach 3.0. I worried that an unstart on my airplane at this point would make it difficult to catch up and perform my chase duties for the Coldwall test. They said that would be fine, and gave priority to the inlet data. This highlighted the low priority of the chase mission. So, I flew through the wake of the YF-12A as soon as we reached Mach 3.0, passing within a hundred feet of the other airplane. The shock wave pattern streaming off of 935 was extremely strong, and the YF-12C responded as if a giant hand had grasped it and twisted it back and forth. I was very impressed

NASA photo EC75-04775

YF-12A (top) and YF-12C (SR-71A) in flight during a heat transfer research mission. The exposed "Coldwall" experiment package is visible beneath the fuselage of the YF-12A.

with my airplane's inlet system. The spikes and bypass doors did a dance, making very large movements, and I worked hard to keep the aircraft level. Much to my surprise, the inlets never unstarted.

NASA photo E75-28414

After the YF-12A's ventral fin broke off during a flight, NASA search teams located it in the desert. Here, Mallick props up the fin as other search team members look on.

A Few Add-on Tests Along the Way

Occasionally, flight research yielded some interesting phenomenon entirely by chance. This often led to the addition of previously unscheduled experiments onto a program. Once, when Bill Campbell and Sam Ursini were performing some sideslip evaluations in the YF-12A, the leading edge of the lower ventral fin was bent sideways by aerodynamic loads. This was unexpected, and the airplane flew without a ventral fin for a time while Lockheed technicians repaired the damage. Loss of the ventral fin during the Coldwall program further indicated that we needed to make some changes.

The ventral fin had failed at both hinged attach points and the engineers studied the flight data to discover why it happened. They decided that the "center of pressure" on the ventral was located much further aft than had been expected for the flight conditions at the time of the incident. The center of pressure is a point on the surface of an airfoil where all of the various aerodynamic forces would be centered if they were all somehow summed together. It is somewhat analogous to the center of gravity, a point where all weight would act on an object if applied at one point without disturbing the balance of that object. The center of pressure is normally located about one third of the way back from the leading edge in subsonic flight and about halfway back in supersonic flight. The loads data indicated that the center of pressure was much further aft than expected at Mach 0.95, creating higher bending loads and resulting in the unexpected failure.

We searched the desert at the foot of the Sierra Nevada mountains for the lost ventral fin. It took about a week to find it and bring it back to Edwards. It wasn't in bad shape, all things considered. Minor damage from ground impact made it impractical to repair and use again. Since we needed to replace the fin, we used the opportunity to test a new material. Lockheed manufactured the new fin out of a substance called Lockalloy. Short for "Lockheed Alloy," it consisted of 62 percent beryllium and 38 percent aluminum. In the meantime, we scheduled more flights for stability and control data to evaluate the YF-12's directional stability without a ventral. Based on our data, the engineers concluded that the ventral was not as critical to adequate directional stability as the designers had thought. This unanticipated test program provided worthwhile data and was typical of the way such efforts sometimes evolved at NASA FRC.

Other Areas of Investigation

During the course of the YF-12 program, we conducted many types of experiments and investigations. One study sought to improve autopilot function for aircraft like the proposed supersonic transport (SST). We tested cockpit improvements such as the presentation of inertial altitude rate and inertial longitudinal acceleration on the pilot's instrument panel. This made it easier to fly precise propulsion test points. The engineers attached structural panels of various materials to parts of the aircraft to evaluate heating and loads effects. On most flights, we also gathered medical information. The flight crews were always wired with "biomed" sensors during test flights to provide physiological and biomedical research data on pilot workload and stress.

Fitz and I both found that on certain missions, even though we had flown the Blackbird recently and considered ourselves up to speed, we had difficulty establishing our test points, and the time required to establish each point was much longer than normal. NASA researchers who tracked the atmospheric data taken during our flights discovered that we were encountering unstable air masses. On some days, the temperature at a given altitude varied a great deal over a relatively short period. These thermal variations changed the Mach number of the aircraft, sometimes up to 0.1 Mach. We needed to fly our test points to within 0.05 Mach, so this phenomenon made difficult to establish a test point. On days when we had a stable atmosphere, the aircraft remained at a given Mach number for minutes at a time. As a pilot, I was greatly relieved to have an explanation for the problem. I think a good research engineer or pilot always wants an explanation and is never satisfied to accept a phenomenon as unknown.

Fitz and I worked closely together and flew a lot of the same research programs at NASA for over 20 years. We certainly knew each other's flying abilities and characteristics. We were a "well oiled" crew and good friends. Since we shared so many programs together, we also carpooled most of those years. Over the span of my career, I have flown with many outstanding Navy, Air Force, and NASA pilots. I have always been reluctant to say that any particular one was the best, but I can't think of one that was better than Fitz. On top of all his flying talent, he was a fine "Southern Gentleman" with a "super cool head." I know that I became a better pilot from flying programs with Fitz. It's like the golfer that becomes a better player when he plays with tougher competition. Fitz was tough competition, not that it was anything but a friendly rivalry. He was a "tough act to follow," and he inspired me to strive for excellence. He was also a good manager. On numerous occasions, I sought his opinion and drew on his experience to assist me in my position as chief pilot and deputy director of operations. Fitz and I occasionally socialized after work, but not as much as you might expect. I think that was due to our many hours together at work, and the fact, that we both needed some "away time" at home.

10

A Time of Transition

The End of an Era

The 1970s were a turbulent time. Many events remain etched in my memory of that era. The moon flights continued to provide human drama as my old friend Fred Haise helped bring the damaged Apollo 13 spacecraft back to Earth after a very dangerous mission. Apollo astronauts scored triumph after triumph, collecting scientific data until the final lunar journey in December 1972. A former U.S. president, old "Give 'em Hell" Harry Truman, died a week after the Apollo 17 crew splashed down.

In 1971, FRC started a program to study a supercritical wing configuration that would promote fuel conservation through improved aerodynamics. It was eventually incorporated into many civilian and military aircraft designs. The following year, we installed a digital fly-by-wire (DFBW) control system in a modified F-8C Crusader. DFBW controls are also now common in modern aircraft and spacecraft.

In April 1973, NASA suffered its worst aircraft accident to date. Seven NASA and four contractor personnel died when the Convair CV-990 flying laboratory from Ames Research Center collided with a Navy P-3 on approach to Moffett Field near San Francisco. It was the same airplane I had flown for the aurora research missions. Four of the five Navy P-3 crewmen also perished. The sole survivor was nearly killed by one of the fire trucks that responded to the scene.

On 31 October 1973, astronaut Ken Mattingly gave me a ride to Houston in a T-38 so I could pick up a Bell 47G helicopter. The next day, I began a four-day journey back to Edwards in the tiny helicopter. The overall distance of my route was about 1,500 miles. I made it in 25.5 flight hours with nine refueling stops along the way. With an average ground speed of 52 miles per hour, I was happy when I could keep up with traffic on the highways.

On 26 March 1976, NASA FRC became the NASA Hugh L. Dryden Flight Research Center in honor of a pioneering aeronautical scientist and former director of the NACA. A month later, the first Space Shuttle orbiter, named "Enterprise," rolled out of the Rockwell International facility in Palmdale. It signaled a new approach to space flight, featuring a reusable vehicle that launches like a rocket and lands like a conventional airplane.

In the spring of 1977, the Office of Aeronautics and Space Technology (OAST) Research Council held a meeting to discuss upcoming programs. The

Council consisted of all the center directors and some program managers from NASA Headquarters. Our center director was playing political games to obtain what we called "new start" money for new programs. He offered to terminate the Blackbird program, thinking that the other center directors who had interests in the project would refuse this, thus giving him the new start money and allowing the YF-12 research to continue. The gamble backfired, and the Council decided to phase out the Blackbirds by 1979. The decision was generally unpopular and, in my opinion, very stupid. There were programs lined up waiting for the aircraft through the mid-1980s. I never thought very much of that particular center director's management style. He was gone from NASA Dryden by the end of October, but the damage had been done. The YF-12 research program ended with a final NASA flight on 31 October 1979. It is interesting to note that in 1991, NASA Dryden began a research program involving several SR-71 Blackbirds. It was a great loss that so many years went by without having these aircraft available for high-speed flight research. This resulted from a political decision that was not based on good research planning.

The LLRV, XB-70, and YF-12 were my "big three" NASA research programs. The YF-12 was my last major, high-visibility, prime research project. Before, during, and after these programs, I was involved with numerous other worthwhile and interesting research efforts. Although they had less national visibility, they also made significant contributions to the development of aerospace technology.

Paying the Toll

Over the span of the Blackbird program, I increasingly noticed the effects of the physical demands of flying. All of the pilots at NASA Dryden were aware of the importance of being in good physical condition. NASA provided a workout facility for us in the 1960s, and we took advantage of it. For years, the Dryden pilots went to the Lovelace Clinic in Albuquerque, New Mexico, for an annual physical examination that included a large number of stress and aging tests designed especially for aviators. Almost all of the pilots quit smoking and were concerned about their physical shape and longevity for long-term test flying. The medical doctors and researchers at Lovelace encouraged this and tracked our progress closely year after year, noting any degradation of physical or mental performance.

At this time, there was no maximum age restriction for NASA pilots. We were able to continue flying as long as we were physically qualified. In the military, the career ladder required that pilots move on to other areas, including command and management. By the time an Air Force pilot reached age 45, he had usually left high-performance flying behind in favor of a command staff position. I started to fly the Blackbirds at age 40 and finished at age 50. In the last year or two of the program, the pressure suit flights took more out of me.

NASA photo EC76-05270

Mallick and several other NASA pilots flew the Northrop YF-17 for base drag studies and to evaluate the maneuvering capabilities and limitations of the aircraft.

I felt more fatigue after each flight. In the early 1980s, NASA Headquarters established an age limit of 61 years for piloting single-seat, high-performance aircraft. Representatives of the various flight operations groups within NASA helped establish the guidelines. Even though some of the operations leaders from the various centers didn't like to add limitations, they agreed that it was reasonable to do so.

Northrop YF-17

In the summer of 1976, I had an opportunity to check out in the Northrop YF-17. Only two YF-17 prototypes were built for a "fly-off" competition sponsored by the Air Force in the early 1970s. The twin-engine YF-17 squared off against the single-engine General Dynamics YF-16. Ultimately, General Dynamics won the contract to produce the next lightweight, multirole fighter for the Air Force. The Navy apparently watched the competition with interest and saw promise in the YF-17. It later served as a prototype for the Navy's F/A-18 Hornet.

NASA Dryden arranged to borrow the first YF-17 prototype for base drag studies, and to evaluate the maneuvering capabilities and limitations of the aircraft. I was one of seven NASA pilots selected to fly the YF-17 and had the

NASA photo EC76-05706

Mallick poses with the YF-17 following a flight in the summer of 1976.

The Smell of Kerosene

good fortune to make two qualitative test flights in the airplane. This was a state-of-the-art fighter jet with very impressive performance. I found the flight control system fairly "user friendly" and the flying qualities excellent. "If this airplane lost the competition," I thought after flying the YF-17, "the YF-16 must be a fine aircraft indeed."

B-52 Mothership

Because of my B-52 experience in the 1960s, I was assigned to operate the NB-52B mothership, along with Fitz Fulton. Originally, starting with the X-15 program, the Air Force provided the aircraft, maintenance personnel, and flight crews. It also served as a launch platform for the heavyweight lifting bodies and other projects. As the joint NASA-USAF lifting body operations began slowing down, the Air Force decided that they were no longer going to provide the support to NASA. In April 1976, they transferred custody of the NB-52B to NASA on permanent loan.

In October 1973, we started an air-launch program with the Remotely Piloted Research Vehicle (RPRV). In fact, the RPRV was a 3/8-scale model of a McDonnell Douglas F-15A fighter plane. It had no propulsion, but included a complete flight control system. We carried the F-15 RPRV aloft under the wing of the NB-52B and dropped it from 50,000 feet altitude. A pilot controlled it remotely from a ground station that featured a simulated cockpit. We used the RPRV to gather stall and spin data. The subscale F-15 was completely instrumented and we collected data through telemetry and ground-based optical tracking cameras.

McDonnell Douglas Aircraft Company built three of the vehicles for the relatively low price of $250,000 each (compared to $6.8 million for a full-size F-15). Initial flights were recovered in mid-air by helicopters, but later flights employed skids for horizontal landings on the lakebed. In 1977, the F-15 RPRV was redesignated the Spin Research Vehicle (SRV) and modified to evaluate the effects of various nose configurations on the stall/spin regime. On some flights we had time to induce two complete spins and execute recoveries prior to landing.

On one of the early flights, the helicopter had snagged the RPRV and was headed back to base when trouble struck. Tom McMurty, who had remotely piloted the F-15 model, sat back in his ground-based cockpit to relax after his portion of the flight was completed. Suddenly, he noticed on his television monitor that the cable holding the RPRV had broken. The F-15 model fell, inverted, toward the desert below. Tom quickly grabbed the controls, rolled the vehicle upright, and made a nice emergency landing on the Edwards bombing range. It all happened very quickly. The cable broke when the aircraft were less than 1,000 feet above the ground. Tom was praised for his quick reflexes and smooth response under pressure.

This turned out to be the final use of the mid-air recovery system (MARS) for the RPRV. On subsequent flights, the vehicle landed on the lakebed. It was

NASA photo EC73-03804

The NB-52B served as the "mothership" for many experimental vehicles. Here it is seen with the remotely-piloted Spin Research Vehicle (SRV).

just as well. The MARS required the use of Navy and Air Force helicopters, which made logistics more complicated. Additionally, MARS recoveries were not always successful. Failures to engage the recovery parachute resulted in increased hazards to the helicopter crews and damage to the RPRV. The skid system eliminated these hazards and also gave the NASA ground pilots experience in making remotely controlled landings.

Long before the Space Shuttle ever launched, NASA Dryden personnel helped develop recovery parachutes for the spacecraft's solid rocket boosters (SRB). During launch, two SRBs help propel the Shuttle towards orbit. Once the booster rockets burn out, they are jettisoned and fall back to the ocean under recovery parachutes. A special boat picks them up and tows them back to Cape Canaveral for inspection and refurbishment for future flights.

To test the proposed parachute system, we used a parachute test vehicle (PTV) that simulated the weight of a real SRB. The SRB PTV was carried to altitude by the B-52, using the old X-15 pylon on the right wing. Since it weighed about 50,000 pounds, extra fuel was loaded into the left wing to balance the B-52 aircraft for takeoff. Once the B-52 was safely airborne, much of this lateral balance fuel was transferred back to the fuselage tanks. The pilots used aileron trim to maintain aircraft attitude while flying to the launch point with about one-half of the lateral static unbalance that the SRB PTV normally caused. At launch, the unbalance shifted to the other side, but it was easy to handle with the ailerons and spoilers. When the PTV departed the aircraft, there was a loud metallic "crack" as it cleared the pylon and the B-52 pitched up and rolled to the left. I found it easy to handle these forces with normal control inputs.

Things didn't always work smoothly, and one event stands out in my mind. Fitz was in the left seat and the controls for the electric-hydraulic release mechanism were on his side. As we approached the launch point over the Edwards Precision Impact Range Area (PIRA), Fitz activated the release switch, but the PTV failed to drop. We spent about ten minutes flying a racetrack pattern back to the drop point and discussed the problem with NASA controllers. They decided we should attempt to drop the PTV with our backup emergency release system. It used air pressure to open the hooks holding the PTV in place. When we reached the drop point, Fitz pulled the manual release handle and fired the source-air bottles located in the nose. We could hear air hissing, but the PTV remained stubbornly in place. If that wasn't bad enough, we now had a red light on the launch panel showing that the test vehicle was unsafe on the hooks. This meant that the hooks had started their movement but did not go far enough to release the PTV. We were now between the proverbial "rock and a hard place."

We were faced with the possibility of landing the B-52 with the PTV attached, but in a partial-release condition. If it released during touchdown, the 25-ton PTV could roll under the landing gear. The results would be catastrophic. Fitz and I discussed our options. We decided not to deploy the drag chute at the normal speed after touchdown. It had a tendency to jerk the aircraft, which might release the PTV. We planned a long landing roll, using every bit of the 15,000-foot-long concrete runway. We planned to use the drag chute once we slowed down enough that the force of its deployment would be minimized. I called the tower and declared an emergency landing, with a request that fire trucks be located near the runway.

As we turned onto final approach, I thought to myself that if I had to be involved in a problem like this, there was no one that I would prefer to have with me than Fitz. I had flown the B-52 with him for years and I couldn't recall him ever making a rough landing. I knew that, had I been in the left seat, I would have requested that he take the controls and make the landing. I flew the B-52 well, but this was a case for the man with the most experience. Fitz was definitely that man. He made a beautiful, smooth, wings-level touchdown within 500 feet of the approach end of the runway. He gently deployed the wing spoilers to add drag and came in ever so gently on the brakes. As our speed dropped somewhat below the normal chute speed, he called for the drag chute, which I deployed. At the lower speed, the chute caused only a gentle

A Time of Transition

tug as we slowed down. Then, as we like to say in the postflight debrief, we made an "uneventful rollout."

Ground support personnel approached the aircraft to insert safety pins into the launch mechanism, but they would not fit. Fitz and I shut the engines down on a parking area just off of the west end of the runway. It was time for the maintenance crews to work on the problem. We had finished our part and I was damned glad. A postflight investigation revealed corrosion of the launch mechanism. Fitz and I were just glad that it hadn't dropped off during the landing roll. When I had a chance to mentally review the potential for disaster, I realized that hazards like this were part of the job. I had used up another silver bullet this day. During the landing, there wasn't time to dwell on the threat. I just did everything that was required to handle the emergency and relied on God and good luck to get me through.

In 1979, the Air Force asked NASA Dryden for help in improving the parachute recovery system for the F-111 crew capsule. The F-111 fighter/bomber had a two-place cockpit, with the crew seated side by side. Instead of individual ejection seats, the airplane had an escape capsule. The cockpit separated, deployed several parachutes, and floated to the ground intact. It had been used a number of times, with mixed results. To test the improved parachutes, we dropped a new kind of PTV from the bomb bay of the B-52 at various speeds and altitudes. The test article looked like a big red box with a pointed front end. It was weighted to simulate the F-111 escape capsule. Over a period of several years, we conducted a number of test series for the Air Force as they developed different parachute systems and staging scenarios.

Low-speed drops were never a problem, but high-speed drops often provided a challenge. As usual, we planned our drops for stabilized test conditions. That was easier said than done. When we opened the bomb bay doors about two minutes prior to drop, the old B-52 wanted to do everything but stabilize. As the doors opened, the airplane pitched up and yawed sideways. The additional drag from the open bomb bay caused the B-52 to slow down. Fitz and I were busy, working as a team to control the beast and stabilize on the planned drop speed and course.

Also in 1979, we started using the B-52 to launch the Highly Maneuverable Aircraft Technology (HiMAT) RPRV. Rockwell International Corporation built two HiMAT vehicles as low-cost, high-performance technology demonstrators. The subscale airplanes simulated advanced fighter designs with sharply swept wings, canards, and winglets. They were built from the most modern alloys and composites, similar to what might be found in a new combat aircraft. A ground pilot flew the HiMAT remotely, as with the F-15 RPRV. This time, however, the RPRV was powered by a jet engine and capable of transonic speeds and complex maneuvers.

The Air Force co-sponsored the HiMAT program with NASA. The organizations involved hoped to validate design tools with flight test data. At

NASA photo EC82-21064

NASA used a Martin B-57B for wind-shear and atmospheric turbulence studies

the time, it was the most sophisticated RPRV ever flown. By flying such an advanced design remotely, there was no risk to the pilot. Critics still argued that RPRV system lacked the flexibility inherent in manned flight. Nevertheless, NASA Dryden and HiMAT proved that the RPRV technique was a useful tool for flight research. The two HiMAT vehicles made 26 flights between July 1979 and January 1983. They provided data on aircraft structures, aerodynamics, flight controls, and propulsion.

B-57 Atmospheric Studies

During the early 1970s, I participated in wind-shear and turbulence studies using a B-57B. Several airliners had crashed due to encounters with powerful downdrafts close to the ground. These "downbursts" were usually associated with thunderstorms. Because Edwards was not prone to stormy weather, I flew the B-57 to other parts of the country for the tests. The area around Denver, Colorado, was of great interest due to its history of downburst activity. The goal of these gust gradient evaluations was to learn as much as possible about the nature of downbursts in order to prevent future accidents.

The flight crew for weather research missions consisted of a pilot and a flight test engineer (FTE). We based our search for downburst activity on information from Doppler radar coverage of the Denver area that tracked and identified wind shear. We tried to fly the B-57 through the downburst as many times as we could during it's "lifespan," which might be a few seconds or up to 20 minutes. Locating a downburst and collecting useful data was a tremendous challenge. Once, I found a classic example right at the end of a mission when my fuel was low and I was preparing to land. I decided that I was going to get some data, even if I had to deadstick the B-57 into Buckley Airport. We penetrated the downburst at an altitude of 1,500 feet and 180 knots. This speed gave us a safety factor. In the event that we hit severe turbulence, we would not exceed the aircraft's structural load limits. Our goal was to hold our altitude and speed with engine power. In other words, as the downburst started to drive us toward the ground, we added more and more engine power to remain on track. During this particular downburst, I went to maximum power and was still losing altitude. I held the speed at 180 knots and flew out of the other side of the downburst 400 feet lower than I had entered it. The diameter of the downburst was initially small, about half a mile. As it "aged," the diameter spread to four or five miles. I could clearly see dust blowing on the ground as the high velocity winds reached the surface, spread outward, and boiled up, providing a visual indication of the event's circumference. At any rate, I got some good data and landed safely.

Our gust gradient research continued sporadically into the early 1980s. Fitz and I shared piloting duty in the B-57. Our FTEs included Vic Horton, Ray Young, and Marta Bohn-Meyer. On occasions when ground-based radar did not offer any useful information, we flew around, hunting downburst activity near thunderstorms and cumulus clouds that formed in the spring and summer. On one occasion, Marta and I spotted two large cumulus build-ups east of Denver, just beyond the Doppler radar coverage area. We alerted ground control that we were going in to investigate. This was a slight deviation from our normal operating procedures. The ground controllers liked to have the subject storm that we were investigating on their radar in order to have correlation of airborne and ground data, and to alert us to the possibility of hail or other dangers.

We flew toward the two storm cells, both of which appeared quite active. I flew a circle around both cells. They were close together, with a small clear area between them. Not detecting any downdrafts, I decided to make a pass between the cells. As we approached the mid-point, a tremendous bolt of blue-white lightning flashed in front of the aircraft. The flash momentarily blinded me and my hair was standing on end. I felt static electricity inside the cockpit and I smelled ozone. When I regained my composure, I made a tight 180-degree turn and proceeded back the way I had entered. After deciding that the airplane and crew were O.K., I began to wonder if the lightning had damaged any of our sensitive instrumentation. Fortunately, it had not and we proceeded home. I have always had a great respect for thunderstorms and my lightning

The Smell of Kerosene

NASA photo EC83-24364

Don Mallick and Marta Bohn-Meyer used the B-57 to gather data on "downburst" phenomena. Don flew the airplane and Marta served as flight Test Engineer as the B-57 penetrated the powerful downdrafts.

experience strengthened that respect. I just wanted to get a little data, but those two cells didn't want to give it up that day.

Lockheed JetStar

The longest research aircraft program that I was involved with was the JetStar. NASA purchased a Lockheed L-1329 JetStar (we also called it by its military designation, C-140) in 1963. I flew that airplane until 1986, just a year before my retirement. Both of us came to Edwards at about the same time, but the airplane retired before I did. I recall joking to co-workers that it was time for me to go because one of my favorite aircraft had finally been put out to pasture.

Initially, the JetStar was converted to a General Purpose Airborne Simulator (GPAS) with wide capabilities and general application. Modifications to the aircraft were extensive, and cost more than NASA's purchase price of around one million dollars. The airplane had good performance and flying qualities, but more importantly, it also had a reasonable amount of space to install the hybrid analog-digital computer that was the heart of the simulator system. NASA technicians installed special servo-motors on all flight control surfaces and a fly-by-wire system with a unique control "feel" mechanism on the pilot's side. The co-pilot's side was connected to the original aircraft controls and was the back-up aircraft control system in the event that the computer or associated equipment malfunctioned. A simple emergency disconnect switch released the aircraft from all of the fancy computers and returned it to basic control from the right seat. For this reason, we referred to the right seat position as the "safety pilot" and the left seat as the "evaluator pilot."

Three companies bid on the contract to convert the JetStar to the GPAS configuration. Cornell Aeronautical Laboratory won the final technical and cost bids for the job. At that time, Cornell had as much aircraft simulator experience as anyone in the country. Initially, the contract review team thought Cornell's bid was too low and that they could not possibly complete the job at the price they quoted. When questioned, Cornell officials emphatically insisted that they could do it. As it turned out, Cornell ran out of money just prior to the end of the contract. NASA subsequently arranged for another $100,000 to complete the task. Fortunately, there was little resistance from the NASA contract office. Cornell was doing an outstanding technical job, but they made the mistake of under-bidding and NASA apparently decided not to punish them. As I learned later, after talking to Cornell management personnel, the company did not make any real carry-over profit from the project. They kept their team working for about 18 months, and they had the satisfaction of producing a fine airborne simulator, but that was it. Despite the money problem, NASA had a lot of respect for Cornell and maintained a long working relationship with the lab.

NASA photo EC82-20293

The JetStar performed numerous research functions during its long career. Here it is seen with a scale model of an unducted fan (UDF) engine mounted on top of the fuselage.

One of the first aircraft simulated with the GPAS in 1965 was the XB-70, which was then undergoing flight test with the Air Force. The computer on the JetStar was programmed with the equations-of-motion and flying characteristics of the XB-70, as they were known from earlier flight tests. Some XB-70 pilots were then invited to fly the JetStar and evaluate the simulation in order to validate it as accurate. We scheduled XB-70 pilots Joe Cotton and Fitz Fulton (who was still with the Air Force at the time) to fly the JetStar on the same day that they had flown the XB-70. With a shorter time interval between flying the two aircraft, the pilots were better able to evaluate of the accuracy of the simulation. As it turned out, the GPAS did a remarkable job of simulating the XB-70's flight characteristics. Eventually, XB-70 pilots used the JetStar to help develop damping techniques to use on the XB-70 in problem areas of poor lateral-directional characteristics with dampers off around Mach 2.3.

In 1971, direct-lift controls (DLC) and side-force generators were installed in the JetStar to enhance its capabilities. Direct-lift was provided by splitting the original landing flaps and installing extra servos to drive the DLC flap. The side-force generators consisted of two large vertical surfaces mounted below the fuselage behind the cockpit. The GPAS computer coordinated inputs to the side-force generators and aircraft rudder to give additional sideward motion to the aircraft. This modification gave the JetStar GPAS more motion capability than most of the aircraft it simulated.

The GPAS was considered to be a low risk research program. If any problem occurred, a simple push of a button put the aircraft out of danger and back to the basic JetStar control configuration. On 14 June 1974, after about nine years of safe operation, we discovered that our reliable, safe JetStar had a surprise waiting for us.

We had been using the direct-lift controls to investigate the gust alleviation capability of the system. Stan Butchart was the safety pilot in the right seat and I was the evaluator pilot in the left seat. We had just completed an evaluation point at 300 knots indicated airspeed and 25,000 feet altitude. At this point, the test card called for disconnecting the GPAS and gust alleviation system. When I disconnected the system, the JetStar immediately started shaking up and down. The sudden onset of the high-frequency oscillation startled us. Stan acted quickly. He pulled the nose up, deployed the speed brake, and pulled back on the throttles. The oscillations gradually slowed. For more than 45 seconds, the JetStar bucked and shook as if it wanted to tear itself apart. Slowly, after what seemed like an eternity, the shaking damped out and we found ourselves still flying. Stan and I finally had a chance to catch our breath. We looked at each other with relief and consternation. Clearly, both of us had the same question: "What in the hell was that?"

We called our control room on the radio, declared an emergency, and said we were returning to base. NASA launched Gary Krier in a T-38 chase plane to join us as we approached Edwards and to look us over for damage. Gary reported seeing wrinkles and bends in the wing tanks and landing flaps. We decided to make a no-flaps landing on Rogers Dry Lake. The tower cleared us for an emergency landing on lakebed runway 18 and we made an "uneventful" touchdown (i.e., a landing with no further damage to the aircraft).

The JetStar was completely instrumented, so there was plenty of recorded data to review. After many hours of work, investigators concluded that the direct-lift system had been designed without a mechanical lock on the flap portion of the DLC system. When the DLC was shut off, only the hydraulic oil in the servos held the flaps locked. When the servos were not operating, they had no hydraulic pressure and only the hydraulic fluid provided a solid link to the flaps. Investigators believed that, on this particular flight, we had some air in the hydraulic system. At 300 knots, the aircraft was in a regime conducive to flutter. As soon as we turned off the DLC (removing the hydraulic pressure), air in the lines allowed the trailing edges of our flaps to move. They immediately began to flutter and sent the wing into its natural frequency.

The JetStar was built tough and it survived this experience only because of its natural strength and Stan's quick response to the emergency. We had three research scientists in the back of the JetStar with us that day running the computers and recorders, and all five of us used up one of our "silver bullets."

NASA photo

The "Front Office." Mallick in the JetStar cockpit.

There are few pilots or crewmembers alive today that have experienced and survived such extreme aerodynamic flutter. It usually results in destruction of the aircraft. When I arrived at home that evening, my wife detected an odd look on my face and asked me if all had gone well that day. I told her about the incident. Even though she was not familiar with the seriousness of aircraft flutter, she was familiar with her husband, and she realized that the "old man" needed a stiff drink of Jack Daniels that night.

The aircraft was grounded for several months to repair the external fuel tanks, fairings, and flaps. We were also concerned about the integrity of the internal wing tanks. One of the original wings had leaked for years until we replaced it. The replacement wing had been obtained from another JetStar that had suffered a ground oxygen fire that ultimately destroyed the fuselage. Fortunately, we didn't have to replace any wings this time. There was no apparent leakage following the flutter incident.

The engineers decided not to re-design the direct-lift control system. The split flaps were locked together with doubler straps, returning them to their original configuration. For years after that, whenever I did my preflight "walk around," I looked at those "strapped" flaps and remembered my wild ride.

In 1976, we conducted a laminar flow control (LFC) study with the JetStar. Technicians modified the wing to improve airflow over the surface. Researchers wished to determine if it was feasible to use enhanced laminar flow control systems on future airliners. The LFC wing had numerous tiny holes in the surface and a device to create suction through the holes. The suction system kept the boundary layer of air across the wing surface laminar (smooth) and prevented it from becoming turbulent, thus reducing aerodynamic drag.

From 1976 to 1982, we also used the JetStar to checkout the Microwave Scanning Beam Landing System (MSBLS) for the Space Shuttle. We flew over 346 missions and 671 flight hours for this program, checking each of the planned primary landing fields for the Shuttle. They included Edwards AFB, Kennedy Space Center in Florida, and Northrup Strip at White Sands, New Mexico (now called White Sands Space Harbor). The designers hoped to develop a fully automatic landing capability for the orbiter, but it was never fully achieved. We used a laser tracker to provide the exact position the JetStar in flight in order to validate the accuracy of the MSBLS. It was extremely accurate, but had one problem that was never conquered. There was a small wave anomaly in the longitudinal microwave signal close to the surface due to the ground effect on the microwaves. We were never able to cancel this wave to make the system acceptable for auto-land. The MSBLS was eventually used on the Shuttle orbiter for the landing approach down to the last several hundred feet, but it was not used for a fully automatic landing. In 1988, when I heard that the Soviets had flown a completely unmanned automatic flight on their shuttle right through landing touchdown, I was surprised. I wondered why we couldn't do that. Later, I learned that they almost lost it during landing, so I didn't feel so bad.

In 1978, we added a working scale model of an ultra-high bypass (UHB) propfan engine to the JetStar. The UHB propfan was designed to improve fuel consumption. Jet exhaust drove two counter-rotating turbines that were directly coupled to the fan blades, eliminating the need for a gearbox to drive a large fan. Our propfan program involved a 24-inch diameter advanced high-speed propeller. The propeller and its air turbine drive motor were installed on the top of the JetStar fuselage with a special pylon. Technicians added an armor plate strap around the fuselage below the propeller in case the experimental propeller shed some blades or disintegrated. Researchers were considering the propfan for application to high-subsonic speed turboprop transports, but they were concerned about the acoustics of supersonic propellers. The JetStar had an array of microphones installed across the aircraft fuselage and wings to record the sound of the fan blades, which were running at supersonic speeds at the propeller tips. During this program, we accomplished 72 missions and 131 flight hours between 1978 and 1982. The JetStar was such a flexible research platform that we had no problem using it for both the propfan and MSBLS programs during the same period.

In 1983, we began a second phase of the LFC program. This involved a more substantial modification to the JetStar than the original GPAS work. Lockheed and McDonnell Douglas corporations each built a test section for the aircraft's wings. Installing these sections on the JetStar required extensive structural work. For budgetary reasons, NASA Dryden completed modifications "in house." It was the largest aircraft modification completed at the NASA facility, and demonstrated the capabilities of Dryden personnel. Some fuel tanks had to be reduced in size to accommodate plumbing for the LFC suction equipment. The vacuum pumps pulled air from the boundary layer over each test section, resulting in extension of the laminar flow back across the airfoil, thus reducing aircraft drag and improving efficiency. Each of the two contractors used a different approach to their wing-test section design, and thus provided more flight data for comparison. Test results indicated that the LFC sections were successful. Next, we wanted to see if they might be applicable to a real-world environment.

We put the JetStar "on the road" and flew it from large major airports around the nation. We flew in all types of weather, including winter snow and ice. The LFC installations continued to perform well. Soot and smoke from jet exhaust at major airports did not degrade the system's performance and our de-icing systems were adequate to keep the laminar flow sections working in winter conditions. The biggest drawback was the size and complexity of the system. This made it expensive to build and install. We proved that LFC worked, but it was simply not economically practical to apply the system to commercial airliners because the cost exceeded the payback.

Other NASA Pilots

A number of other pilots joined the Flight Operations office during my tenure. Their teamwork and dedication helped make NASA Dryden a very special place to work and a great asset to the agency and the nation.

When Fred Haise and I transferred to FRC in 1963, John Manke worked as a flight test engineer on the X-15 program. John was an ex-Marine fighter pilot with F8U Crusader experience. He left the service in 1960 and worked for Honeywell Corporation as a test engineer for two years before joining NASA. He wanted to transfer into Flight Operations. At the time, center director Paul Bikle turned him down. "John," said Bikle, "I can always find pilots, but I can't always find good flight test engineers." In 1965, when Fred Haise joined the astronaut corps, John transferred to the pilot's office to replace Fred. He flew the F-104, T-33, T-38, F5D-1, and C-47 support aircraft, and served as a co-project pilot on the F-111 and general aviation programs. In 1968, he made his first HL-10 flight. He went on to make 42 lifting body flights in the HL-10, X-24A, M2-F3, and X-24B. Later in his career, John became the center director. Our families were very close together during our years at NASA. Audrey and I socialized more with John and his wife Marilynn than with any other NASA friends. John retired from NASA in 1984.

Tom McMurtry, a former Navy pilot, joined NASA in 1967. He was a graduate of the Navy Test Pilot School and also had experience flying operational missions in the U-2. Tom was the first pilot that I hired in my capacity as chief pilot. He began his NASA career flying the T-33 and ended up flying everything from small general aviation airplanes to the Boeing 747 Shuttle Carrier Aircraft (SCA). He earned the NASA Exceptional Service Medal for his work with the F-8 Supercritical Wing (SCW). By the time I retired in 1987, Tom had moved up to head the Aircraft Operations division. He eventually retired as Chief Engineer in 1999.

Gary Krier, a former Air Force pilot, joined NASA in 1967 after working as an engineer for Pratt & Whitney, Martin Marietta, and Hercules Powder Company. Gary was the first pilot to fly the F-8 Digital Fly-By-Wire (DFBW) aircraft and served as co-project pilot on the F-8 SCW. He flew more than 30 types of aircraft ranging from light planes to the B-52. He later earned a law degree and moved on to NASA Headquarters for a management position related to Space Shuttle operations. As much as I hated to lose him, it pleased me to have someone of his caliber in Headquarters management. He returned to NASA Dryden in 1995 and continued to work his way up the management ladder.

Hugh Jackson joined the pilot's office in June 1966. While at NASA FRC, he flew the T-33, C-47, JetStar, F-104, F5D-1, F-4A, B-57, and various light aircraft. In 1967, he had the opportunity to fly an Acme Sierra S-1 Sportplane that Northrop used to demonstrate their concept of a close air support combat aircraft. Hugh was quite intelligent, and earned a doctorate degree in his spare time. I saw him study now and then, and he made an occasional trip back to his school in Illinois. One day in early 1969, he asked me for a few days leave to take his "orals." When he came back, he was Dr. Jackson. Hugh left NASA in October 1971 and became an aviation consultant.

Early in his career, Einar Enevoldson flew high-performance fighter aircraft with the U.S. Air Force. As an exchange officer with the British Royal Air Force (RAF), he graduated from the Empire Test Pilot School and served as a test pilot on RAF Hunter, Lightning, and Javelin fighter planes from 1966 to 1967. Einar joined NASA in January 1968. He flew many aircraft at the center, including the T-33, F5D-1, PA-30, F-111, F-8 DFBW, F-8 SCW, F-14, AD-1 Oblique Wing, X-24B, YF-12A, and the Controlled Deep Stall sailplane project. He also served as a remote pilot for the F-15 RPRV. He was awarded the NASA Exceptional Service Medal twice. He received it first, in 1974, for his work on the F-111 Transonic Aircraft Technology program and the F-15 RPRV and again, in 1980, for his contributions as project pilot on the F-14 stall and spin research project. Einar retired from NASA in 1986 and became chief test pilot for the German Grob Egret long-range, high-altitude, turboprop airplane.

Steve Ishmael, then an active pilot with the Air National Guard, joined the Dryden team in 1977. He was one of the "new breed" of pilots with an extensive background in electronics and computers. He had a Master of Science degree in engineering (with emphasis on computer technology). This was a desirable background for piloting the newer aircraft with their digital control systems. While I was at Dryden, Steve flew the T-37, T-38, F-104, JetStar, PA-30, F-111, C-47, F-16, F-8 DFBW, F-14, and the forward-swept-wing X-29. In 1988, he cheated death by ejecting from a crippled F-18 northeast of Edwards. He served as chief project pilot on the F-16XL Laminar Flow Control and SR-71 High-Speed Research programs before leaving the pilot's office for management duties in 1995. I recall Steve as a highly motivated individual. He was so dedicated to getting a research pilot job that he frequently hung around the office even before the pilot selection was complete. I suspect that Steve did not want us to forget who he was. We didn't. Steve was selected for tasks because he was the most qualified candidate of all the applicants.

Rogers Smith served as a pilot with the Royal Canadian Air Force from 1955 to 1963. He worked as a research pilot at NASA Langley in 1967 before joining the CALSPAN Corporation (formerly Cornell Aeronautical Laboratory), where he flew the X-22A VSTOL testbed and NT-33A Variable Stability Trainer. Rogers came to NASA Dryden in 1982 with an extensive background in airborne simulators, flight control systems, and pilot displays. He flew a wide variety of airplanes including the T-37, T-38, F-104, JetStar, Bell 47G helicopter, F-111, C-47, B-52, F-8 DFBW, B-57, Boeing 720, X-29, and Boeing 747 SCA. He also flew F-4 Phantoms in the Air National Guard. He remained at Dryden after I retired, serving as SR-71 project pilot and eventually being promoted to Chief Pilot. He retired in 1999 to pursue a career as a consultant.

Jim "Smoke" Smolka came to Dryden in 1985 after testing F-16s for the Air Force at Edwards and working on some of the joint USAF/NASA F-16 projects. A graduate of the Air Force Academy and Air Force Test Pilot School, Jim continued to serve with the Air Force Reserve after joining NASA. He flew the F-104, T-38, JetStar, and the F-111 Mission Adaptive Wing testbed. After I retired, Jim went on to fly the F-16XL, B-52, F-18 High Alpha Research Vehicle (HARV) and numerous F-15 research projects.

Ed "Fast Eddie" Schneider joined the Navy in 1968. In 1973, he became the youngest graduate in the history of the Navy Test Pilot School. He remained at the Naval Air Test Center as an engineering test pilot and instructor test pilot, and later served as F-4 program manager and senior production test pilot at the Naval Air Rework Facility at North Island, California. During his last tour of duty, in 1982, Ed was assigned to NASA Dryden as Navy Liaison Officer. This was done to establish closer ties with the Navy. We had a long-running, excellent relationship with the Air Force because we operated from their base and shared a number of joint programs with them. Ed was a super guy, very bright, and one of those individuals who operate at 125 percent all of the time and makes it look normal. Ed helped NASA procure Navy F/A-18 aircraft for Dryden as replacements for our aging F-104 fleet. In 1983, a vacancy opened up in the pilot's office and Ed was encouraged to apply. He did, and

NASA photo EC77-08608

In 1977, NASA Dryden carried out a series of Approach and Landing Tests with the Space Shuttle Enterprise. The prototype orbiter was launched from a modified Boeing 747 and made unpowered landings at Edwards AFB.

was selected. I felt that Ed was an outstanding addition to the Dryden team. He became involved with the F-14 Automatic Rudder Interconnect (ARI) and laminar flow research projects, F-8 DFBW, F-104 aeronautical research and microgravity studies, F-15, B-52, F-18 HARV, Boeing 720 Controlled Impact Demonstration (CID), and SR-71. Ed transferred to Johnson Space Center in September 2000 as a T-38 instructor pilot.

As new pilots came to Dryden, the "old heads" began retiring. This was just normal attrition. Fitz Fulton retired more than a year before I did and I was the last of the highly qualified "heavy" pilots. Several of the newer pilots were flying with me as co-pilot in the B-52, but they were some time away from being promoted to airplane commander. A very fortunate thing then happened.

C. Gordon Fullerton transferred from Johnson Space Center to Dryden. "Gordo" had joined the Air Force pilot in 1958. He flew the F-86 and B-47 and later attended the Aerospace Research Pilot School (now the Air Force Test Pilot School). In 1966, he was selected as a crewmember for the Air Force Manned Orbiting Laboratory (MOL). The MOL was cancelled in 1969 without ever flying in space, but Gordo was assigned to NASA as a support astronaut for the Apollo lunar missions. In 1977, he flew the Space Shuttle Enterprise during the Approach and Landing Test (ALT) project. Gordo subsequently piloted the Space Shuttle Columbia on its third orbital mission in 1982 and commanded Challenger during the STS-51F mission in 1985. In November 1986, Gordo came to NASA Dryden permanently. His flight experience was a great asset to the center. I felt that he came from the same mold as Fitz Fulton. Just the thought of having him in the office made my departure much easier. I had been in the organization a long time and I wanted to leave it in good order. Gordo filled the only void that I could see in the operation, and that was a pilot with extensive experience in large airplanes. I checked Gordo out in the B-52 during three flights and was extremely satisfied with his performance. He went on to fly numerous other projects including the F-111 Mission Adaptive Wing, F-14 Variable Sweep Flow Transition Experiment, JetStar Laminar Flow Control, X-29, Boeing 747 SCA, and numerous projects involving the B-52.

11

Shifting Gears

Another High Flyer

During the course of the YF-12 program, I maintained close ties with the Air Force Blackbird community at Beale AFB, near Marysville, California. We had a mutually beneficial exchange of information regarding our similar operations. Beale, home of the 9th Strategic Reconnaissance Wing, also operated the Lockheed U-2. Two of their aircraft, designated U-2CT, were trainers with an instructor's position above and behind the student cockpit. In July 1981, I had the opportunity to visit Beale for checkout in the U-2CT.

On the morning of 21 July, I flew the NB-52B, with an Air Force co-pilot, during a captive flight of our second HiMAT vehicle. That afternoon, I packed my overnight bag into one of our F-104N support aircraft. John Manke and Vic Horton climbed into a TF-104G and we flew to Beale. The following day we were briefed on the U-2 and prepared for checkout. The U-2 looks like a glider, because of its long wingspan, but it has a powerful turbojet engine. Although it can only fly at subsonic speeds, the U-2 is capable of flying above 70,000 feet. On takeoff, "pogo" wheels kept the wingtips off the ground. The pogos fell away as soon as the U-2 became airborne, but the airplane had skidplates to protect the wingtips during landing.

The next morning, John and I each got to make a flight in a U-2CT with an Air Force instructor. We flew for nearly two-and-a-half hours, and it was a wonderful experience. I really enjoyed flying this exotic bird. Landing was a special challenge because the U-2 had bicycle-type gear. There was a set of two main wheels under the center of the fuselage and two small tail wheels near the aft end. My flight included some no-flaps landings that were very exciting. I had to fly an extremely flat approach and pull the engine power to idle about two miles from the touchdown point. The U-2 then sailed in like a high-performance glider with just a bit of idle thrust from the engine pushing it onward. I had to use careful judgment in order to avoid touching down short of the runway. Since I had flown sailplanes some years earlier, I wasn't completely unfamiliar with this type of performance. That experience helped me a great deal, but it was more difficult landing the U-2, with or without flaps, than it was landing a glider with air brake spoilers. "The U-2," said the Air Force instructors, "separates the men from the boys." I had to agree with them.

Lockheed photo CC-1576

Mallick had the opportunity to fly a Lockheed U-2CT at Beale AFB, California. In some respects, he found it similar to flying a sailplane.

Changes

The biggest change in my career during the 1980s was that I became more involved with management than actual research flying. As deputy of the Flight Operations division, I was now involved in all aspects of the research missions. This included conducting and monitoring various technical and administrative meetings required for the preparation of each flight, and directing the control room during missions. My control room responsibilities were sometimes more complex than actually flying the mission. When I flew, I just made decisions and judgement calls regarding how I operated my aircraft. Whenever I was in charge of the control room, I sometimes made decisions for another pilot, especially regarding safety issues. I tended to be conservative, always sticking to our mission plan. NASA flight rules prohibited adding new test points during a mission. Adding something extra, just because some researcher decided that he would like to see something else on the spur of the moment, was a good way to get into trouble. All mission plans underwent a rigorous review and approval process. Once in awhile, I upset a few people by terminating a flight early due to some equipment failure or when I observed something that made me question the safety of the flight. I didn't mind, because that was my prime

The Space Shuttle Columbia lands at Edwards AFB on 14 April 1981 following the first orbital flight test. The Approach and Landing Test (ALT) program at NASA Dryden paved the way for the orbital missions.

responsibility when I was in charge of the control room. During the mission, I received continuous inputs from highly qualified personnel in different disciplines concerning various systems health and status. This critical information helped me make decisions regarding the safe continuation of the flight. Sometimes, I felt that I would have rather flown the mission myself, but I held a position of greater responsibility now.

On 12 April 1981, the first Space Shuttle orbital test flight took place with John Young and Bob Crippen onboard piloting the orbiter Columbia from Kennedy Space Center (KSC), Florida. They landed at Edwards on 14 April, touching down majestically on Rogers Dry Lake. The Shuttle operation was just beginning, as was Dryden's support of the flights landing at Edwards. The lakebed served as the primary Shuttle landing site until NASA was prepared to begin landing operations at KSC. Even afterward, orbiters continued to land at Edwards whenever bad weather prevented a landing in Florida.

In October, budget cuts forced NASA Headquarters to merge NASA Dryden Flight Research Center with Ames Research Center. As a result, Dryden lost its "Center" status and became a "Facility" subservient to Ames. It was a purely political move that probably could not be avoided. NASA convinced Congress that this was their way of accepting a budget cut without closing a center completely. I supported Headquarters on this one, because it was the better option. The merger did have a long-term negative effect on Dryden, as the facility was now managed from several hundred miles away. I suspect that it actually cost NASA more money to operate in this manner, but it was the politically correct thing to do at the time. Dryden's biggest losses were to our prestige and morale. In spite of the change, Dryden continued its tradition of unique and innovative flight research.

Losses

For some years now, I hadn't lost any close friends to flight accidents. Some test pilots had been killed during these years, but none in my immediate circle and none within NASA. In June 1980, a legendary Lockheed test pilot named Herman "Fish" Salmon died in the crash of a Lockheed Constellation aircraft. The "Connie" was a large, four-engine airliner that had been out of service for years. Fish retired from Lockheed in 1978, but continued checking out crews and ferrying some old Constellations that were being refurbished. Fish was contracted to ferry a Super Constellation to Alaska in June 1980. The aircraft lost power on takeoff from Columbus, Indiana, killing Fish and two other crewmen. I had met Fish when I was on the membership committee of the Society of Experimental Test Pilots (SETP). I recall Fish's weathered skin, "crow tracks" around his eyes, and his laid-back, unassuming attitude. I also recall the humorous stories that he and Tony LeVier would tell about each other. The best flying stories were usually told when pilots got together for a few drinks and relaxation. Often, such tales were self-deprecating. Pilots sometimes seem to enjoy confessing that they are not completely perfect. At times, they had done something that was not especially smart, but they had survived.

On 28 July 1981, one of NASA Dryden's research pilots died in a recreational sailplane accident. Mike Swann, a surprisingly experienced young aviator, had transferred from the Johnson Space Center (JSC) flight operations branch in 1978. At JSC, he had piloted high-altitude Earth resources and air sampling missions in the WB-57F and served as an instructor and check pilot for the Astronaut Space Flight Readiness Program. At Dryden, he was involved in the F-111 Transonic Aircraft Technology (TACT) program, F-8 DFBW, PA-30 Remotely Piloted Research Vehicle study, and numerous other programs. Mike became interested in glider flying and purchased part ownership of a sailplane with Einar Enevoldson. Mike took a day off work to fly his glider and practice some entry maneuvers for a cross-country sailplane competition. The entry maneuver involved passing through a starting "gate," and the idea was to be at high speed to allow him to climb and start on the cross-country flight. Mike was practicing these entries north of the California City airport when the tail of his glider came off. He spun to earth with no chance to bail out. I flew up the next morning in the Bell 47G helicopter and searched for the portions of the tail that were missing from the wreckage. I located them about half a mile north of the impact point. The two fiberglass horizontal stabilizers had separated completely, probably due to aerodynamic flutter. There was no instrumentation on the aircraft, and it could only be assumed that Mike had either exceeded a "red line" (maximum performance) limit of the glider, or that the tail had failed below the "red line" due to a structural weakness. There was no way of knowing which. In any event, we had lost a very promising young pilot. The National Transportation Safety Board (NTSB) handled the accident investigation because Mike was not flying on NASA business.

On 11 September 1981, Ames Research Center lost a small, two-place Hughes OH-6A helicopter when it flew into high-voltage power lines in a foggy mountain pass. Dave Barth was killed and his co-pilot, Richard "Tex" Ritter, was so badly injured that he never returned to flying status. I became quite familiar with the details of this accident, as I was assigned as chairman of the investigation board. It was not unusual for one NASA Center to request assistance from another for accident investigations. I had served on several other accident boards, including an investigation of a LLRV that crashed at JSC. No lives had been lost in those cases, however. Of all the tasks I had performed in my career, this was the most distasteful. I had to come in after a pilot had been killed and put myself inside his mind to see where and why he might have screwed up. The helicopter and engine wreckage showed no signs of failure prior to impact, and a ground witness had observed the helicopter circling before entering the clouds near the power lines. The final report described a sequence of events that was based on all the evidence we reviewed. The board determined that the pilots had intentionally entered an area of reduced visibility at low-altitude, resulting in collision with the power lines. NASA Headquarters gave our report a chilly reception. It seemed that they wanted us to say more directly that the pilots used poor judgment. Well, they obviously had, or they wouldn't have crashed. The members of the board felt that anyone with aircraft knowledge could interpret that from the report.

On 8 November 1982, NASA Dryden lost another new pilot. Dick Gray spun in and crashed in a twin-jet Cessna T-37B trainer. Dick had transferred to Dryden from JSC in November 1981. A former Naval aviator, he had flown 48 combat missions in Vietnam and served as a test pilot at NAS Point Mugu, California. Dick's last Navy assignment was flying the F-14 at NASA Miramar, near San Diego, California. In 1978, he joined NASA and flew the WB-57F at JSC. When Dick came to Dryden, he was assigned to the AD-1 Oblique Wing, F-8 DFBW, and the F-14 Aileron Rudder Interconnect (ARI) programs. In preparation for the ARI project, Dick underwent a special Navy spin training program at NAS Patuxent River, Maryland. When Dick died, he was flying a

NASA photo EC82-17956

Mallick had the opportunity to fly many unusual vehicles during his career, including the Ames-Dryden AD-1 Oblique Wing research aircraft. The Space Shuttle Enterprise is in the background.

local proficiency flight in the T-37 to practice spins. He apparently failed to recover from a flat spin and attempted, unsuccessfully, to eject shortly before impact.

Needless to say, everyone at the Center was shocked to lose two Dryden pilots in such short order. Center management felt that we really needed to step back and review our procedures. Although, strictly speaking, Mike's accident was not related to his NASA duties, we wanted to review our flight operations procedures and look for problems. It was imperative that we prevent further accidents. It is difficult, if not impossible, to determine what goes on in a pilot's mind that results in a pilot-error-type accident. People have been trying to do that for years, but the human being is a complex mechanism. "To err is human," as they say.

Mike and Dick were both were intelligent and aggressive individuals. They were dedicated to excel in their field. They knew that test flying was inherently risky. They were both known for outstanding performance and recognized for their flight test experience. These factors may have led them to overextend their own capabilities and place themselves in danger. We always say that a pilot must "stay within the aircraft envelope," because within that range of performance, the aircraft's behavior is known and predictable. A pilot must also stay within his own performance envelope and not exceed his capabilities. This is true in other areas in life, as well.

NASA flight operations have always relied on the professionalism and the good judgment of the individual pilots to stay within their own envelope and that of the aircraft they are flying. There is no written rule. It happens because the pilots make it happen by their own integrity and self-discipline. Professionalism is expected of each pilot, and rightly so. It was often frustrating when managers from NASA Headquarters approached an accident review board with the attitude that operational management was primarily responsible for the accident. I'm sure there were occasions when this was true. Operational procedures force a pilot to continuously integrate, filter, and react to hundreds of inputs. Many of these can set him up to contribute a "pilot error" or "pilot factor" to an accident. But, the bottom line is that the pilot is the last in the loop and is ultimately responsible for the aircraft.

Dick Gray's accident had one worthwhile result. Dryden Flight Operations initiated a requirement for planning and monitoring of any flights, including proficiency, where unusual and demanding maneuvers were to be performed. We believed that if there had been a few more individuals in the loop and more review, perhaps the spin accident could have been avoided.

Controlled Impact Demonstration

In 1984, NASA teamed up with the Federal Aviation Administration (FAA) for a unique project to improve crash survivability. The NTSB had determined that many airplane crashes would have been survivable if not for the post-impact fire. As the aircraft broke apart, fuel sprayed into the air and ignited, with deadly results. Subsequently, a contractor developed a fuel additive to prevent this problem. The product, called Anti-Misting Kerosene (AMK), showed great promise in laboratory tests. NASA and the FAA proposed to test the properties of AMK using a full-scale transport aircraft and actually crashing it by remote control.

The FAA supplied a Boeing 720B airliner for the project, called the Controlled Impact Demonstration (CID). NASA technicians modified the airplane as a remotely-piloted vehicle (RPV), controlled from a ground cockpit. The airplane was completely instrumented and included cameras inside and outside the fuselage. There were even dummy passengers onboard. Prior to the final CID mission, 14 test flights were made with a four-man safety crew onboard. During these flights, the ground pilot performed 10 remote takeoffs, 13 remote landings, and 69 simulated CID approaches. All remote takeoffs were flown from the Edwards AFB main runway. Remote landings took place on the emergency recovery runway (lakebed runway 25).

The Smell of Kerosene

NASA photo EC84-31806

The Controlled Impact Demonstration employed a remotely piloted Boeing 720 airliner for a crash test. A special fuel additive was supposed to reduce or eliminate a fireball on impact. Sometimes failure teaches more important lessons than success.

The final CID flight, on 1 December 1984, was entirely remote-controlled by Fitz Fulton, with no safety crew onboard. Fitz set up the aircraft's final approach to a specially prepared test area on the east side of Rogers Dry Lake. It included a series of metal fixtures to slice open the wings as the jet crash-landed on the lakebed. As the jet began its slide across the hard-packed clay, it yawed slightly. One of the airplane's engines struck a metal fixture and disintegrated in a ball of fire. A massive fireball quickly enveloped the wing and fuselage as the airplane broke into pieces. It was a stunning failure. Investigators later determined that the AMK had settled in the fuel tanks during the night and failed to mix sufficiently during flight to be effective.

As I have said before, it is often just as important to learn what doesn't work as to learn what does. Critics called it a failure, but we knew better. Flight research and test is the process of validating, or invalidating, hypotheses. By testing AMK under fairly realistic circumstances, we proved it didn't work as advertised. Prior to our demonstration, the concept seemed so promising that the FAA proposed legislation requiring use of the additive in some types of jet aircraft. Our data prevented further investment of time or money into a bad idea.

Challenger

I was onboard an airliner, on 28 January 1986, when I heard the news that the Space Shuttle Challenger had exploded 73 seconds after launch that morning. Even knowing the complexity and risk involved in Shuttle operations, I was shocked by the news. The shuttle commander, Dick Scobee, had been an Air Force test pilot at Edwards and flown a number of research missions at NASA Dryden. I grieved for all the crew, but especially Dick, who I knew best. I can still recall his broad grin when he visited the Dryden pilot's office following the announcement of his selection as an astronaut. He showed great pride in his selection, and I congratulated him heartily. The results of the accident review board were hard to accept. The commission that investigated the accident blamed the Shuttle loss on poor management decisions. Challenger had been launched against the recommendations of knowledgeable technical personnel who insisted that low temperatures that day increased the chance of hot gas leakage around the seals of the solid rocket boosters. The commission found that the decision making process leading to the launch was flawed and that launch temperature constraints were waived at the expense of flight safety. It was a black day for NASA. I could sense a change in people's attitude concerning the space program. After the Challenger accident report was released, the public's pride in and respect for NASA diminished.

At Dryden, we had always striven not to allow the desire to "get a flight off" to interfere with good judgment on flight safety. It was a cardinal rule. There were occasions when visiting Headquarters personnel and other VIPs were on hand to witness a test flight and we had to cancel the event due to some technical problem. We forced ourselves to avoid the desire to "press on" just to meet a schedule or impress a visiting VIP.

Final Thoughts

I have always been proud of my association with NASA. I experienced a burst of pride whenever I was introduced to someone as a NASA employee. Like most government agencies, NASA has experienced political difficulties and other problems, but it has also accomplished a lot of extraordinary things.

I could not have found a better flying job for a test pilot than the one I found with the NACA and NASA. The people that I worked with were "The Best of the Best." The agency's philosophy toward flight research was one of high integrity. A demand for gathering important data was balanced against a high degree of concern for flight safety. I can never recall making a flight that I considered unsafe in any way. I can't recall anyone in NASA ever asking me to put myself or an aircraft at risk.

I flew a wide variety of aircraft over the years. The opportunities at NASA Dryden were almost unlimited. Dryden was sometimes referred to as NASA's "Skunk Works," meaning that we were able to get things done without an excess of paperwork and wasted time or money. "Skunk Works," a term borrowed from Lockheed, has come to mean a small, innovative operation that produces impressive results.

I must confess that, in the later years, the paperwork caught up with us. It is impossible to run such a complex operation without extensive documentation. I felt that there was both good paperwork and bad paperwork. The good paperwork included documentation required for the safe and efficient conduct of flight operations. Bad paperwork consisted of excessive documentation for the sole purpose of "covering your ass" or CYA. Sometimes, the documentation requirements that had been established in the name of safety actually reduced safety. Keeping up with documentation required so much of the manager's valuable time, that it was a distraction. That time could have been better spent concentrating on efforts more directly involved with improving safety. Such misuse of time bothered me greatly, and the hours that I spent responding to those CYA requirements were given grudgingly.

During my career, I saw both triumph and tragedy. I hope that I have not over-emphasized the latter at the expense of the former. Research and test flying is a hazardous profession, although not as much now as in years past. It is also a very challenging and rewarding profession. I feel very fortunate to have had the opportunity to be a test pilot and experience life on the cutting edge of aerospace technology. I would like to thank all of my associates for their support and contributions to my career. I was but one tooth on a large gear that was a part of the aeronautics development machinery in this country, for about 30 years. I hope my successors will enjoy the same productivity and good fortune that I have. I offer a special final salute to those very good friends I knew who were killed during their careers. I have great hopes that there is the "world after," and that one day I will share a drink and some good stories with them all again.

A Final Whiff of Kerosene

On 3 April 1987, I made my final NASA flight. The morning was cool and overcast. Raindrops quickly soaked into the parched desert landscape. As I strapped myself into the cockpit of the F-104N, many thoughts passed through my mind. With bittersweet nostalgia, I worked through my preflight checklist and requested clearance to taxi. I eased the throttle forward gently and began the three-mile trip to the Edwards main runway. A few drops of rain streaked my canopy, but the sky was clearing. It would be a beautiful day.

Upon reaching the runway, I made final checks and received takeoff clearance. I accelerated down the 15,000-foot concrete strip, lit the afterburner, and climbed toward the heavens. For one glorious hour, I danced among the clouds. Finally, I returned to Earth to end one chapter of my life and begin another.

Back on the Dryden ramp, I shut the engine down and opened the canopy. Sunlight dappled the wet concrete where steam rose in thin streamers. As the wind shifted, I caught the familiar scent of burnt kerosene.

Shifting Gears

NASA photo E73-25788

F-104N number 811 streaks over the desert between the clouds. Mallick ended his NASA career with a flight in 811 on 3 April 1987.

Epilogue

By the time I retired, I had logged over 11,000 flight hours in more than 125 types of aircraft. My wife and I remained in the Antelope Valley, the area over which I had conducted most of my research flying. Retirement gave me a chance to spend more time with my family, something I welcomed. I also continued to participate actively in my community. Seasonally, I prepare income tax forms under the AARP-IRS Volunteer Program and have served as a Board member with the Antelope Valley Committee on Aging (AVCOA). AVCOA supervises the Meals On Wheels Program, Dial-A-Ride, and other services for senior citizens.

I have also remained in close touch with my friends and colleagues in NASA and other parts of the flight test community. In 1992, while visiting Dryden, Tom McMurtry asked if I would like to fly one of the F-18s that replaced NASA's aging F-104 fleet. Of course, I said yes. Ed Schneider, as instructor pilot, took the back seat of a two-place TF-18. We flew over the Antelope Valley for a little over an hour, and it was just like old times. Ed showed his confidence in me by not once touching the controls. Racing through the azure sky, my grin was hidden beneath the oxygen mask, but the joy in my heart was reflected in my eyes. I was in my element.

Glossary

Angle-of-Attack (AOA) – The angle between the chord of a wing's profile (median line) and the direction of movement, i.e. the angle at which the wind strikes the airfoil.

Area Rule Principle – A method of reducing transonic and supersonic drag on an airplane by narrowing the fuselage cross-section at the center of the fuselage. Such tapering gives the fuselage a "wasp waist" or "Coke bottle" shape.

Arresting Cable – A metal cable stretched across the deck of an aircraft carrier to halt a landing airplane. The airplane snags the cable with a hook lowered from the aft fuselage.

Boundary Layer – As a fluid (such as air) moves past an object, the molecules right next to the surface stick to the surface. The molecules just above the surface are slowed down by their collisions with the molecules sticking to the surface. These molecules in turn slow down the flow just above them. The farther one moves away from the surface, the fewer the collisions affected by the object surface. This creates a thin layer of fluid near the surface in which the velocity changes from zero at the surface to the free stream value away from the surface. Engineers call this layer the boundary layer because it occurs on the boundary of the fluid.

Bypass Ratio – ratio of air mass going through a jet engine's bypass system to the total air mass going through the engine's compressor face.

Davis Barrier – A net to catch an airplane that is in danger of overrunning the deck of an aircraft carrier, if it fails to snag the arresting cable.

Delta Wing – An airfoil with a triangular planform. Delta wing airplanes normally do not have horizontal tail surfaces.

Digital Fly-By-Wire (DFBW) – Electronic flight control system in which the cockpit controls are not mechanically linked to the aerodynamic surfaces, but are used only as the commands generator for the automatic flight controls.

Dutch Roll – A three-axis oscillation characterized by a side-to-side wallowing motion.

General Aviation – Pertaining to airplanes owned by private citizens or privately owned companies, as opposed to commercial air carriers or military aviation.

High Alpha – High angle-of-attack.

High Range – High Altitude Continuous Tracking Radar Range. A specially instrumented high-speed flight corridor with several data-linked tracking stations.

Hypersonic – Pertaining to that which occurs at speeds above Mach 5.0 (3,418 mph).

Knots Equivalent Airspeed (KEAS) – Airspeed measured in nautical miles per hour.

Laminar Flow – Boundary layer airflow in which the streamwise velocity changes uniformly as the distance from the surface increases.

Lateral-Directional Stability – Stability characteristics of an airplane that involve the coupling of rudder and aileron inputs affecting roll and heading modes.

Lifting Body – A wingless vehicle that derives stability and lift from the shape of its fuselage alone.

Limit Cycle Oscillation (LCO) – A high-frequency resonant vibration originating in the aft part of an airplane. It can ultimately drive the horizontal stabilizers to their limit-of-travel.

Mach Number – The ratio between the speed an object and the speed of sound at a given altitude. The speed of sound (called Mach 1.0) varies from about 760 mph (1,225 km/h) in warm air at sea level to about 661 mph (1,060 km/h) in very cold air at high altitude.

Military Power – Maximum "dry" engine power, i.e. without use of afterburner.

Model-Following System – Simulation of a linear description (simplified model) of an aircraft's dynamic properties and control system.

Pilot-Induced Oscillation (PIO) – Oscillation caused by the delay between a pilot's control input and the aircraft's response, causing the pilot to make too strong an input or repeat the original input. Trying to correct for the delayed response with opposite input can cause the corrections to be out of synchronization with the aircraft's movements and aggravate the undesired changes in attitude.

Propfan – Ultra-high-bypass ratio turbofan engine combining characteristics of a turbofan and turbo prop engine. Also see Unducted Fan.

Stability Augmentation System (SAS) – An electronic system designed to improve handling qualities.

Supercritical Wing (SCW) – An airfoil with a shape that reduces the effect of transonic shock waves on the upper surface, reducing drag and leading to increased range, speed, and fuel efficiency.

Supersonic – Faster than the speed of sound.
Pertaining to that which occurs within the range of speed from Mach 1.0 to Mach 4.9 (760 mph to 3.205 mph).

Tactical Air Navigation (TACAN) – A ground-based radio navigational aid which provides bearing and distance data to the pilot.

Transonic –
Pertaining to that which occurs within the range of speed in which flow patterns change from subsonic to supersonic (about Mach 0.8 to Mach 1.2) or vice versa.

Unducted Fan (UDF) – A propfan engine without a fan duct. The curved fan blades are exposed to the air like the blades of a propeller.

Visual Omni-Range (VOR) – A high-frequency ground-based radio transmitter used by pilots to determine azimuth information for navigation purposes.

The Smell of Kerosene

Appendices

Appendix A

U.S. Navy Aircraft Flown (1950-1963)

North American Aviation SNJ *Texan* – Basic flight training and basic carrier qualification

Grumman F6F *Hellcat* – Advanced flight training and advanced carrier qualification

Beech SNB-2 *Navigator* – Instrument flight training

Lockheed TO-2/T-33 *Shooting Star* – Jet transition and training

Lockheed TO-1/F-80 *Shooting Star* – Jet training and proficiency

McDonnell F2H-2 *Banshee* – Operational jet squadron service

Vought F4U *Corsair* – Naval Reserve duty, fighter

Grumman F9F-6 *Cougar* – Naval Reserve duty, fighter

Grumman S2F *Tracker* – Naval Reserve duty, anti-submarine

Douglas R4D *Skytrain* – Naval Reserve duty, cargo

Douglas R5D *Skytrain* – Naval Reserve duty, cargo

Douglas A4D *Skyhawk* – Naval Reserve duty, attack

Appendix B

Flight Research Programs At Langley (1957-1963)

Quantitative stability and control (S&C) and handling qualities (HQ) research involving modified helicopters:
Bell H-13G *Sioux* (52-7834)
Hiller YH-32 *Hornet* (55-4968 and 55-4970)
Sikorsky HO3S-1 *Dragonfly* (122520/NACA 201)
Sikorsky HRS-1 *Chickasaw* (127783)
Sikorsky HR2S-1/CH-37 *Mojave* (140316)

Qualitative evaluation of V/STOL aircraft:
Vertol VZ-2 (56-6943)

Developmental test of a "g-limiter" system:
McDonnell F2H-1 *Banshee* (122530/NACA 214)

Aircraft structural dynamics and flutter:
North American Aviation JF-86D *Sabre* (50-509/NASA 205)

Quantitative and qualitative evaluation of flight controls:
Grumman F9F-2 *Panther* (122560/NACA 215)

Qualitative evaluation:
Grumman F11F-1 *Tiger* (138623)

Variable stability flying qualities:
North American Aviation JF-100C *Super Sabre* (54-2024)

Sonic boom studies:
Vought F8U-3 *Super Crusader* (146340)

Support flying, executive transport, and photo chase:
Convair T-29B/CV-240 *Samaritan* (51-7909/NASA 250)
Douglas C-47B/DC-3 *Dakota* (43-49526)
Grumman JRF-5 *Goose* (87748/NACA 202)
Lockheed Model 12A *Electra* (NACA 99)
Lockheed TV-2/T-33 *Shooting Star* (131878/NASA 224)
McDonnell F2H-3 *Banshee* (126300/NACA 210)

Appendix C

Aircraft Flown While At DFRC (1963-1987)

Research Aircraft (Jet):
Ames-Dryden AD-1
Lockheed NT-33A Variable Stability Trainer
Lockheed YF-12A
Lockheed SR-71A/YF-12C
Lockheed U-2CT

Research Aircraft (Other):
Bell Lunar Landing Research Vehicle (LLRV)
NASA M2-F1 Lifting Body

Fighters (Jet):
Convair F-106 *Delta Dart*
Douglas F5D-1 *Skylancer*
General Dynamics F-111A
Ling Temco Vought F-8A *Crusader*
Lockheed F-104 *Starfighter*
McDonnell Douglas A-4 *Skyhawk*
McDonnell Douglas F-15 *Eagle*
North American Aviation A-5 *Vigilante*
North American Aviation F-100 *Super Sabre*
Northrop YF-17 *Cobra*

Bombers (Jet):
Boeing B-52 *Stratofortress*
Convair B-58 *Hustler*
Martin B-57 *Canberra*
North American Aviation XB-70 *Valkyrie*

Trainers (Jet):
Cessna T-37
Lockheed T-33 *Shooting Star*
Northrop T-38 *Talon*

Transports (Jet):
Boeing 707-80
Boeing 727

Boeing 747
Boeing KC-135 *Stratotanker*
Convair CV-990 *Coronado*
Grumman G-1159A *Gulfstream II*
Learjet Model 23 *Continental*
Lockheed L-1011 *TriStar*
Lockheed L1329/C-140 *JetStar*

Transports (Propeller-Driven):
Beechcraft C-12A *King Air*
Douglas C-47
Grumman G-159 *Gulfstream I*

Civil Aviation:
American AA-1 *Yankee*
Beechcraft C33 *Debonair*
Beechcraft D-125
Beechcraft S35 *Bonanza*
Beechcraft UC-45 *Expediter*
Cessna Model 150
Cessna Model 210 *Centurion*
Cessna Model 310
Piper PA-23 *Apache*
Piper PA-24 *Commanche*
Piper PA-28 *Cherokee*
Piper PA-30 *Twin Commanche*
Rockwell Model 680F *Aero Commander*
Wing D-1 *Derringer*

Rotary Wing (Helicopters):
Bell Model 47
Bell UH-1/Model 204 *Iroquois*
Bell Model 206B *Jet Ranger*
Bell OH-58/Model 558 *Kiowa*
Boeing Vertol YH-46A *Chinook*
Sikorsky SH-3 *Sea King*
Sikorsky CH-53 *Sea Stallion*

Sailplanes:
Eiri-Avion PIK-20E
Schweizer SW-126
Schweizer SW-222

Appendix D

Published Technical Papers

Flight investigation of pilot's ability to control an airplane having positive and negative static longitudinal stability coupled with various effective lift-curve slopes
NASA-TN-D-211
Roy F. Brissenden, William L. Alford, and Donald L. Mallick
February 1960

A flight investigation of an acceleration restrictor
NASA-TN-D-241
Arthur Assadourian and Donald L. Mallick
April 1960

Crew performance on a lunar-mission simulation (Included in: A report on the research and technological problems of manned rotating spacecraft)
NASA TN-D-1504
Joseph S. Algranti, Donald L. Mallick, and Howard G. Hatch Jr.
August 1962

Lunar-mission simulation to determine crew performance on a prolonged space flight
IAS Paper 63-18
Donald L. Mallick and Harold E. Ream
January 1963

Crew performance during real-time lunar mission simulation
NASA TN-D-2447
Joseph S. Algranti, Howard G. Hatch Jr., Donald L. Mallick, Harold E. Ream, and Glen W. Stinnett Jr.
September 1964

Flight results with a non-aerodynamic, variable stability, flying platform
SETP Technical Review, Vol. 8, No. 2
Emil E. Kleuver, Donald L. Mallick, and Gene J. Matranga
September 1966

An assessment of ground and flight simulators for the examination of manned lunar landing
AIAA Paper 67-238
Emil E. Kleuver, Donald L. Mallick, and Gene J. Matranga
February 1967

Flight results obtained with a non-aerodynamic, variable stability, flying platform
NASA TM-X-59039
Donald L. Mallick, Emil E. Kleuver, and Gene J. Matranga
September 1968

Flight crew preparation and training for the operation of large supersonic aircraft
Presented at FAUSST VII Meeting, Paris, France
D. L. Mallick and F. L. Fulton, Jr.
March 1969

Recent flight test results on minimum longitudinal handling qualities for transport aircraft
Presented at FAUSST VIII Meeting, Washington, D.C.
Herman A. Rediess, Donald L. Mallick, and Donald T. Berry
January 1971

Handling qualities aspects of NASA YF-12 flight experience
Donald T. Berry, Donald L. Mallick, and Glenn B. Gilyard
Proceedings of SCAR Conference, Part 1 (N77-17996 09-01)
January 1976

Index

Adams, Mike (X-15 accident) 160-161
Adkins, Jim, 139
airfoil model tests, 74-75
airplanes:
    A-5 Vigilante, 104
    A-12, 168, 172
    AD Skyraider, 23, 30-32
    AD-1 oblique wing, 226
    A4D Skyhawk, 94-95
    Aero Commander, 104
    AF-2 Guardian, 41
    AT-6 Texan, 7
    B-26 (NB-26H), 117
    B-29 Superfortress, 97
    B-52B (NB-52B) Stratofortress, 111-114, 204-205, 206-207, 221
    B-52D Stratofortress, 149-150
    B-57, 208-209, 210-211
    B-58 Hustler, 141-142, 147-148, 149, 156-157, 158
    B-70 (XB-70) Valkyrie, 119, (accident) 132-136, 137-138, 139-140, 141-145, 146-147, 149-151, 152-154, 155-156, 157-158, 168, 170, 172-174, 177, 179, 182-183, 201, 212
    Boeing 707, 93,
    Boeing 720, 227-228
    Buccaneer (Blackburn NA.39), 90
    C-47 Skytrain (Gooney Bird), 46, 50, 104, 114-115
    C-140 JetStar, 101, 103-104, 105, 211-212, 213-214, 215-216
    CV-990, 162-163, 164-166, 167, (accident) 200
    FH-1, 21
    F2H Banshee, 20-22, 24, 27, 30, 32, 37, 58-59, 66
    F4D Skyray, 104
    F-4H Phantom II, 81
    F4U Corsair, 20, 23, 39-40
    F5D Skylancer, 104
    F6F Hellcat, 14, 16, 20, 23
    F7U Cutlass, 68
    F8F Bearcat, 14
    F8U-1 Crusader, 67-69, 81
    F8U-3 Super Crusader, 79-82, 83-86, 87-89, 90, 92, 95
    F-8 Digital Fly-By-Wire (DFBW), 200

Index

F8F Bearcat, 14
F9F Cougar, 39-41, 58, 79
F9F Panther, 20, 58, 65-66
F11F Tiger, 58, 66, 67, 92
F-15 Eagle, 204
F-15 Remotely Piloted Research Vehicle (RPRV), 204-205
F-17, *see YF-17*
F-18 (F/A-18A) Hornet, 202, 232
F-80 Shooting Star, 17
F-86, 35, (JF-86D) 58, 77-78
F-100 (JF-100C) Super Sabre, 58, 63, 64, 104-106, 107-108
F-101 Voodoo, 58, 92
F-102 Delta Dagger, 51
F-104 Starfighter, 104, 108-109, 112-114, 132-137, 138-142, 153, 173, 179-180, 182, 194, 221, 230-231, 232
F-106 Delta Dart, 51
F-111, 207
Harrier, 55-56
Highly Maneuverable Aircraft Technology (HiMAT), 207-208, 221
HL-10 lifting body, 168
JRF-5 Goose, 46, 47-51
KC-135 Stratotanker, 179, 184-185, 191, 193
Meteor, 34
M2-F1 lifting body, 101, 104, 114-118, 132, 159
M2-F2 lifting body, 102, 159
M2-F3 lifting body, 168
MiG-15, 35-37
MiG-25, 168
Model 12A Electra, 46, 50
P-47 Thunderbolt, 64
P-51, 41
PA-23 Apache, 120
Schweizer sailplanes, 104
Sea Fury, 32
S2F Tracker, 79
SNB-2 Navigator, 16
SNJ Texan, 7-9, 10-12, 13-14
SR-71, 168, 172, 174-175, (as "YF-12C") 186, 189-191, 192, 201
T-29 (CV-240), 72-73
T-33, 17, 74-75, 77, 104, (NT-33A) 105
T-37, 104, (spin accident) 225-226
T-38 Talon, 200

245

TO-1, 17-19, 22
TO-2, 17
U-2, 221-222
Vampire, 34
VZ-2, 55
X-1 (XS-1), 97
X-15, 72, 98-102, 104, 107, 109-111, 112-115, 119, 132, 144, (accident) 160-161, (Last flight)168, (leftover funds) 170, (High Range) 192, (mothership) 204-205
X-20, 71
X-24A lifting body, 168
YB-49, 98
YF-12, 168-169, 170-171, 172-177, 178-187, 188-191, 192-196, 197-199, 201, 221
YF-16, 202, 204
YF-17, 202-204
Albrecht, William (Bill), 110
Alford, William (Bill), 45, 47-49, 79-83, 87-88, (death of) 90-91, 143
Aldrin, Edwin (Buzz), 131
Algranti, Joseph (Joe), 77, 122, 130
Ames Research Center, 104, 162-163, 168, 200, (consolidation) 223-224, 225
Anti-Misting Kerosene (AMK), 227-228
Armstrong, Johnny, 110
Armstrong, Neil, 99, 131, (LLRV accident) 161-162
Aurora Borealis (Northern Lights), 162-167

Barth, Dave, 225
Bellman, Donald (Don), 136, 139, 141
Bikle, Paul, 117, 120-121, 144-145, 216
Bohn-Meyer, Marta, 209-210
Brady, Tommy, 100
Briegleb, William (Gus), 114
Brown, Harold, 142
Brown, Porter (Porty), 60-61
Bureau of Aeronautics Reserve Training Unit (BARTU), 81
Butchart, Stanley (Stan), 101, 103, 144

Cain, James B. (Jim), 20
Campbell, William (Bill), 170, 174, 197
Cate, Albert, 142-143
Central Intelligence Agency (CIA), 172
centrifuge (NAS Johnsville), 61, 62

Index

Champine, Robert (Bob), 45, 55-57, 67, 69, 79, 91
Cliff, Jack, 15-16
Coldwall experiment (YF-12), 193-196
Conrad, John, 82-83
Controlled Impact Demonstration (CID), 227-228
Cornell Aeronautical Laboratory (CALSPAN), 105, 117, 211, 218
Corum family, 97
Cotton, Joseph (Joe), 138, 142-143, 146, 150-152, 156-157, 212
Cross, Carl, 134, 138-139
Curtis, Billy, 172, 186

Dana, William (Bill), 101-102, 107
Direct Lift Controls (DLC), 212-213
downbursts, 208-211
Drake, Hubert (Jake), 122
Dravo Corporation, 2, 4
Drinkwater, Fred, 163, 167
Dryden Flight Research Center, 97-99, 100-162, 168-232, (origins) 97-98, (initial employment with) 96, 99, (naming of) 200

Eakle, Burke, 6, 21
Edwards, Glen, 98
Enevoldson, Einar, 194, 217, 224
Exercise MARINER, 30-32

Faget, Maxime (Max), 71, 75
Federal Aviation Administration (FAA), 92-94, 119, 135, 184, 192, 227-228
Fritz, John, 138, 142
Fullerton, Charles Gordon (Gordo), 219-220
Fulton, Fitzhugh (Fitz), 111, 144, 146-147, 150, 154-155, 156-158, 162-163, 170-171, 173, 179-180, 190, 194-195, 198-199, 206-207, 212, 219, 228

General Aviation Test Program, 119-121
General Purpose Airborne Simulator (GPAS), see C-140 JetStar
Gentry, Jerauld (Jerry), 115
Glenn, John, 62
Gough, Mel, 42
Gray, Richard (Dick), 225-227

Haise, Fred, 99, 102, 116, 216
Hamilton, Bob (Hambone), 25
Heidelbaugh, Gary, 172

Heinemann, Ed, 95
helicopters:
    Bell Model 47G, 133, 200, 225
    H-19, 52
    HO3S-1 Dragonfly, 56, 57
    HRS-1 Chickasaw, 52
    HR2S-1 Mojave, 54
    OH-6A Cayuse, 225
    YH-32 Hornet, 54
helicopter training, 53-54
High Temperature Loads Laboratory (HTLL), 189-190
High Range, 192
Hoag, Pete, 138
Horner, Richard, 92
Horton, Victor (Vic), 170-171, 190, 195, 209, 221

incidents:
    drop malfunction (SRB PTV), 206-207
    emergency landing at NAS Fallon (YF-12), 192-193
    flutter damage (JetStar), 213-214
    helicopter crash (HO3S-1), 56-57
    iced taxiway (JRF-5), 50-51
    landing gear fire (B-58), 156-158
    lost ventral fin (YF-12), 194, 197
    missing fuel cap (F2H), 32-33
    multiple unstarts (YF-12), 195-196
    uncontrolled roll (LLRV), 128-129
Ishmael, Stephen (Steve), 217-218

Jackson, Hugh, 217
Johnson, Clarence (Kelly), 172
Johnson, Lyndon, 143

Kid, Desmond (Des), 21
Kluever, Emil (Jack), 122, 130
Korea, 34-35
Kraft, Christopher (Chris), 68-69
Krier, Gary, 213, 217

Lacy, Clay, 142
Laird, Dean (Diz), 17-19
Laminar Flow Control (LFC), 215-216

Index

Langley Memorial Aeronautical Laboratory, 42-44, 45-96, (at Muroc) 97
Layton, Ronald (Jack), 170, 186
LeVier, Tony, 224
Lewis Research Center, 190
Little Joe rocket, 75, 76, 77
Lockheed Skunk Works, 172, 229-230
Lovelace Clinic, 201
Ludi, Leroy, 38, 42
Lunar Landing Research Vehicle (LLRV), 54, 119, 122-123, 124, 125, 133, 161, 201
Lunar Landing Training Vehicle (LLTV), 122, 130-131, 162

Mallick (Waite), Audrey (wife), 6, 15-16, 78, 95-96, 102-103, 130
Mallick, Bob (brother), 2-3, 5
Mallick, David Glenn (son), 96, 103
Mallick, Darren (son), 130
Mallick, Donald Karl (son), 78, 103
Mallick, Sandra (daughter), 32, 103
Manke, John, 216, 221
Mattingly, Ken, 200
McClosky, Black Mac, 30
McCollom, John, 142-143
McKay, Jim, 100-101
McKay, John (Jack), 100-101, 109, 144
McMurtry, Thomas (Tom), 217, 232
Mercury project, 70-71, 75
Microwave Scanning Beam Landing System (MSBLS), 215
Miller, Bill, 42
Minnegerode, Don, 21
Model Following System simulator, 105
Muroc, 97

National Advisory Committee for Aeronautics (NACA), (employment with) 42-43, (creation of) 43
National Aeronautics and Space Administration (NASA), origin of, 62, 70
Navy service:
    advanced flight training, 13-16
    basic (pre-flight) training, 6-8
    carrier qualification, 11-13, (advanced) 16, (night) 28-29
    Dilbert Dunker, 7
    enlistment, 5
    flight training, 8-19

    instrument training, 16
    jet transition, 17
    Reserves, 38-41
    solo flight, 10-11
Office of Aeronautics and Space Technology (OAST), 200-201
Osborne, Jim, 21

Parachute Test Vehicle (PTV), (F-111 crew escape capsule) 207, (simulated Solid Rocket Booster) 206-207
Paresev, 115-116
Pennsylvania State College (Penn State), 4-5
Pensacola, Florida, 5-6
Petersen, Forrest (Pete), 109
Peterson, Bruce, 102, 109, (M2-F2 accident) 159-160
Pilotless Aircraft Rocket Division (PARD), 48, 74
Purser, Paul, 75

Ream, Harold (Bud), 77, 122, 130
Reeder, Jack, 42-45, 54-55, 69, 79, 91, 96
Ritter, Richard (Tex), 225
Rogallo, Francis, 115
Rogers, Joseph (Joe), 170
Roman, Jim (Doc), 133-135

Salmon, Herman (Fish), 224
Schmidt, Joe, 83
Schneider, Edward (Fast Eddie), 218-219
Scobee, Francis (Dick), 229
Shepard, Van, 154
ships:
    H.M.C.S. *Magnificent*, 30, 32
    H.M.S. *Eagle*, 30
    U.S.S. *Bennington* (CV-20), 30
    U.S.S. *Coral Sea* (CVB-43), 23, 29
    U.S.S. *Midway* (CVB-41), 23
    U.S.S. *Monterey* (CVL-26), 13
    U.S.S. *Franklin D. Roosevelt* (CVB-42), 22, 23, 24
    U.S.S. *Redfin* (SSR-272)
    U.S.S. *Wasp* (CV-18), 30, 32, 35-36
Slater, Hugh (Slip), 170
Smith, James, 142
Smith, Rogers, 218

Index

Solid Rocket Booster Parachute Test Vehicle (SRB PTV), 205-207
Spin Research Vehicle (F-15 RPRV/SRV), 204-205
Sommer, Robert (Bob), 45, 79, 90, 94-95
Sorlie, Donald (Don), 141
Soulé, Hartley, 50-51, 57
Southeast Asia, 36-37
Space Shuttle, (Enterprise) 200, (Columbia) 223, (Challenger) 229
Space Task Group, 61, 68, 71, 83
Sputnik, 70
Squadrons (Navy and Naval Reserve):
    ASW squadron, 79-80
    VA-861, 94
    VF-171, 27
    VF-172 "The Blue Bolts," 20
    VF-741, 39
    VR squadron, 81
    Sturmthal, Emil (Ted), 146, 156
    Supersonic Transport (SST), 147, 190, 198
    Swann, Michael (Mike), 224-225

Test Pilot School, USAF (TPS), 116-119
Thompson, Milton (Milt), 101, 118, 159
Truman, Harry, 200

Ultra-High Bypass Turbofan (UHB) engine, 215
Ursini, Sammel (Sam), 172, 175, 183, 197

Vensel, Joseph (Joe), 133, 144
Vertical Motion Simulator (VMS), 60-62
VGH recorder experiment, 92-94

Waite, Ron, 110
Walker, Grace, 143
Walker, Joseph (Joe), 96, 99-100, 109, 111-114, 116, 117, 119, 122, 126-127, 128-130, (mid-air collision) 132-143, 146
Wallops Station (Wallops Island), 46, 48, 49 ill., 72, 74-76, 83, 85
weightlessness simulator, 62-63
White, Alvin (Al), 134, 138
White, Robert (Bob), 98, 109
Whitney, Rear Admiral, 37
Whitten, James (Jim), 45, 51, 90, 91
Wiley, Jim, 6

Williams, Walter (Walt), 97-98, 122
Woolams, Jack, 97

Yeager, Charles (Chuck), 117-*118*, 119, 145
Young, William (Ray), 170-*171*, 192-193, 195, 209

About the Authors

Donald L. Mallick joined the National Advisory Committee for Aeronautics as a research pilot in 1957. He retired in 1987 as Deputy Chief of the Aircraft Operations Division at NASA Dryden Flight Research Center. During his distinguished career, Mallick logged over 11,000 flight hours in more than 125 different types of aircraft. He is a Fellow of the Society of Experimental Test Pilots.

Peter W. Merlin has worked as an archivist in the NASA Dryden Flight Research Center History Office since June 1997. He has published *Mach 3+: NASA/USAF YF-12 Flight Research, 1969-1979* (NASA SP-2001-4525) as well as many articles on aerospace history.

Printed in Great Britain
by Amazon